The Big Bang and Relative Immortality

The Big Bang and Relative Immortality

Seminal Essays on the Creation of the Universe and the Advent of Biological Immortality

Sebastian Sisti

Algora Publishing
New York

Library of Congress Cataloging-in-Publication Data —

Sisti, Sebastian.
The big bang and relative immortality : seminal essays on the creation of the universe and the advent of biological immortality / Sebastian Sisti.
p. cm.
ISBN 978-0-87586-606-2 (trade paper : alk. paper) — ISBN 978-0-87586-607-9 (hard cover: alk. paper) — ISBN 978-0-87586-608-6 (ebook: alk. paper) 1. Cosmology, Ancient. 2. Immortality. 3. Immortalism. I. Title.

BD495.S57 2008
113—dc22

2008017167

Front Cover: Big Bang Theory: Synthesized Image Taken By Satellite Cobe
Image: © Jeffrey Markowitz/Sygma/Corbis
Photographer: Jeffrey Markowitz
Date Photographed: April 1, 1992

Printed in the United States

Table of Contents

Chapter 1. Big Bangism

Every now and then theoretical physics stubs its toe against the principle of continuity and when it does, philosophy winces. One egregious example was the theory of continuous creation which postulated that new particles were continuously coming into being, either out of the void or out of something else. In an extraordinary display of intellectual courage, the physicists promoting continuous creation later abandoned it in favor of the Big Bang, which, from the view of the principle of continuity, was like hopping from one ontological quagmire into another.

The principle of continuity may be stated as follows: (a) nothing cannot become something; (b) something cannot become nothing; (c) something cannot become something else.

The Ancient Greek Philosophers

The principle of continuity needs to be distinguished from the theory of continuous existence which is traceable back to Thales (?640–547 BC), the earliest of Greek philosophers. Thales believed that the transformations of water revealed the continuity of existence and came to the conclusion that water was the ultimate nature of all things. Parmenides of Elea (540–480 BC) picked up from Thales and probably made the strongest case yet for continuous existence. So tight was his perception of reality he could find no room in it for empty space; a position which led him to deny the reality of motion. Parmenides is perhaps the most distinguished member of the Eleatic school of philosophy. It was the Eleatics who hammered out the first two legs of the principle of continuity. "Nothing can arise out of nothing," the Eleatics believed, "and no thing can be reduced to nothing." Parmenides' chief opponent among the Eleatics was Heraclitus (?540–475 BC), generally regarded as the father of the principle of change. We never step into the same river twice, wrote Heraclitus; everything is in flux; the world began in fire and will end in fire.

Leucippus (?500–430 BC) introduced the notion of empty space into Parmenides' rigid reality and his disciple, Democritus (?460–370 BC), the father of physics, made everything move again. Democritus agreed with the Eleatics that nothing can arise from nothing and that nothing can be reduced to nothing but, like them, offered no proof. The truth of the relationship of something to nothing and of nothing to something seemed so self-evident it was taken to be axiomatic — which is a pity, because the attempt to prove that nothing cannot become something and that something cannot become nothing reveals the third axiom, which is that something cannot become something else. The first two axioms state that existence is continuous; the third proves that things which exist cannot change.

Teetering on a Two-Legged Stool

The lack of proof that nothing cannot become something and that something cannot become nothing has left us teetering, from the time of the early Greeks, on a two-legged scientific–philosophical stool simultaneously swearing by a law of continuity and a law of change. Darwin's monumental insight, for example, sways between a relentless law of biological change and an equally tenacious accumulation of those changes. Mendel's laws reveal that what appears as change is a mask which succeeding generations reveal as the independence and permanence of genes. Gravity pulls (Newton) or pushes (Einstein), but electromagnetism thumbs its nose at both theories.

The Second Law Denies the First Law

The battle comes to a head in the laws of thermodynamics. The first law confirms the law of continuity. The second law confirms an irreversible natural process in which heat energy flows from one state (hot) to another state (cold), the inexorable outcome of which is the heat death of the universe. In the interim, suns burn themselves out, chairs never grow new, and balls don't bounce of their own accord (though angry marlins and basketball players do). The universe certainly seems to be going in one direction — downhill — and while many thermodynamicists concede that life, through reproduction and growth, defies the second law, life is seen as wholly dependent on the sun's heat that must inevitably obey the second law.

Heraclitan scenarios based on the second law are really contingent on the evidence that the universe is expanding. Since expansion is based on the evidence of galaxies flying outward at nearly the speed of light, the looming question is whether or not gravity will save the day by checking it. Einstein wasn't sure that gravity had that kind of muscle. The failure of gravity to halt expansion would condemn every last particle in every last atom to an infinite joyless ride in which the sputtering light of quarks and electrons vainly attempts to heat the cold, black reaches of the void. How a void might absorb heat and how, therefore, the *ultimate* particle would *lose* heat go unexplained.

The vision is one of hot particles of dust scattered to infinity in the final waves of heat, uncollectible and non-refusable because the void lacks a heat pump. The possibility that Democritus might be right — that motion might be an inseparable property of the ultimate particle; that heat might therefore be indissolubly contained in the ultimate particle; that motion might therefore be not unidirectional but directionless; that the ultimate particle might therefore refuse under its own steam — are filed under x for metaphysix. Philosophical stuff.

Running Down of Universe Implies a Perfect State of Beginning

Every century, or half century in this case, has its cosmological fad. If it is flawed, it eventually goes away. The flaw in second-law cosmology is not in its eschatology (where we're all staring) but in its genesis. Irreversibility implies a beginning of energy since the alternative is to believe that the universe has been running down infinitely backward in time but that infinite time hasn't been quite long enough to achieve maximum entropy. And not just a beginning of energy but a perfect state beginning, because the second law begs the question: running down from what? Was energy born rotting or born with the potential to become rotten?

By implying a beginning of energy, the second law denies the first. It implies that energy not only can be created; it had to have been created. If there had been no evidence of red shift, suggesting expansion, and no evidence of lingering sound waves suggesting an ancient explosion, theoretical physics would have needed a Big Bang theory anyway. It is an unavoidable necessity of the second law. Disorder implies order, a perfect state.

Expansion and Contraction Does Not Deny First Law

Not that all Big Bang theories violate the first law or the principle of continuity. Big Bang theories which see the universe as an infinite series of expansions and contractions violate neither the law nor the principle. The Big Bang theory which violates both is the one which postulates that all the stuff of the universe existed in the form of a pinpoint dot which a gazillionth of a second later exploded with a bang.

There are three glaring implications in the popping proposition. One, the pinpoint dot popped out of the same stuff as itself, which denies a beginning. Two, the pinpoint dot popped out of nothing (vacuum genesis). Three, the pinpoint dot popped out of a third order, i.e., an order other than itself and other than nothing. Implication number two says that nothing can become something. Implication number three says that something can become something else, i.e., something completely different from itself, but not nothing.

What Nothing Means, Or Should Mean

For nothing to become something requires that nothing contain a potential to become something. If it contained a potential (of whatever nature), it would not amount to what we mean by nothing. What we mean by nothing amounts to the absolute absence of all possibility. To argue that we can't possibly know all possibility, that we can't be certain therefore whether or not the void really amounts to a void, is first, fatuous; second, self-contradictory; and third, a denial of the experimental evidence of empty space (Michelson–Morley). Rather than confound the meanings of nothing and something, logic permits us to set up what we mean by a "vacuum genesis" by stating the possibility that a spatially-infinite third order might be filling, or might have filled, what we think of as a void: an order other than the ether. We would then have a primary hypothesis which might be stated as follows:

> Fifteen billion years ago, a third order — let's call it Protogenesis — converted a piece of itself, or all of itself, into the pinpoint dot of the Big Bang. Except for the result of its conversion (our universe), we have no way of knowing what Protogenesis might have been or might be; it is indeterminate; absolutely and totally beyond knowability. Despite the insistence on unknowability, however, there are all sorts of certainties buried in the primary hypothesis.

The Folly of a Protogenesis

For example, we can be certain that if a piece of Protogenesis converted itself into our universe, the piece had to have been absolutely separate from the whole of Protogenesis, or the entirety of the whole would have had to convert when the piece converted. Unless they were absolutely separate, we would have to say that they were fused, which would entail the conversion of the whole. To insist that the piece was a piece, i.e., a broken-off part of the whole of Protogenesis, raises the question of whether or not they were absolutely identical as well as absolutely separate. Denying that they were absolutely identical would deny that the piece was really a piece of Proto.

Conversion and Timing Factors

On the other hand, if we concede that the piece was, indeed, absolutely identical with Proto, we have to acknowledge that the piece had to have contained the same conversion factor as the rest of Protogenesis, as well as the same timing factor. While we can't know the true nature of the conversion and timing factors, our primary hypothesis says that Protogenesis converted so, to the degree that the hypothesis is correct, we can infer a conversion factor to Protogenesis. Since there was an implicit delay between the instant in which Protogenesis remained Protogenesis and the instant in which it converted itself, we are free to infer a timing factor: something which triggered the ability to convert at a particular moment.

To argue then that a piece of Protogenesis, absolutely separate from but absolutely identical with Protogenesis, converted itself into the pinpoint dot of the Big Bang would require that they both had to have converted simultaneously; or it would require that we deny that the piece was absolutely identical with the whole. To insist, for example, that the timing factor in the piece of Protogenesis was slightly different from the timing factor in the whole of Protogenesis would prevent us from assuming that the piece was a part of the whole. It would also present us with the sticky problem of the two slightly different timing factors which would have triggered conversions billions of years (or minutes) apart.

No Parts, Only Wholes

The need for the piece to be absolutely identical with the whole also requires that they be absolutely identical in size, in shape, in every property and attribute. We would have two Protogeneses, absolutely identical in size, shape, properties and attributes, both of which had to have converted at precisely the same moment into our single pinpoint dot. Both would have had to disappear.

Absurd as the notion of two absolutely identical Protogeneses may seem (one is certainly unnecessary), the absurdity is traceable to the mistaken assumption that there can be parts of wholes. Reality doesn't give us any parts; only wholes. Either two entities are absolute identities or they are absolute opposites. Either way they would have to be absolutely separate which would make them absolute wholes. An absolute whole may be any fancy our imaginations permit but it can never be part of anything. So we're stuck with two Protogeneses, or one Protogenesis, which hypothetically converted themselves/itself in the single pinpoint dot of the Big Bang.

Remembering that there might have been two, let's stay with the more reasonable possibility of one and consider whether or not Protogenesis was spatially infinite or spatially finite. If Protogenesis had been spatially infinite, it would have had to suck itself into something less than spatial infinity before, or as, it converted; otherwise its conversion would have resulted in a spatially-infinite universe (hardly expandable) instead of a pinpoint dot. What mocks that possibility is that a spatially-infinite Protogenesis sucking itself in from the never-ending reaches of infinity would still be at it. If there were a Protogenesis at all, it would have had to have been spatially finite.

If It's Finite, It Has an Edge

The problem with a spatially-finite Protogenesis is that finitude implies limit to what we call extension, which we attempt to define with words like length, breadth, and depth, which we then attempt to quantify with numerical values. While the spatial finitude of Protogenesis helps us to gain three new certainties about it (length, breadth, depth), we can't be certain that something is finite unless we can say that it ends here, here, and here, in effect describing its edge. Unless we can describe the edge of something — saying that it ends here, here, and here —we are forced to follow its wholeness to infinity, thereby denying that it is finite. The problem is twofold: (1) we cannot describe the edge of something without the ability to say with certainty that it ends here, here, and here; (2) we can't say that something ends here, here, and here unless it is suspended within a void, or nothing.

For example we could argue that the edge of a spatially-finite Protogenesis was extended within still another order called Protoprotogenesis. Theoretically, we'd be setting up two different orders. But to be two different orders they would have to be absolutely separate. The only way they can be absolutely separate is to be separated by an absolute void — nothing — empty space. The gap of empty space could be as fine a line as we might imagine but without it we would have to say that the two orders were absolutely fused and we would have to deny that they were two different orders. Absolute separation allows us to discover their edges and to say that one ends here, here, here while the other ends there, there, there. On the other hand, if they were absolutely fused we could find no edges and we would have to say that they were one and the same and not two different orders.

Einstein's universe is a perfect example. Einstein believed that the universe is spatially finite. But unless Einstein's universe is suspended within an absolute void, it would be impossible to hold that it is spatially finite. At the outermost reaches of the universe's gravitational field, we would have to be able to say that this is the edge of the universe; it ends here, here, and here. If we were unable to find that edge, we would be forced to search outward to infinity and therefore deny that the universe is finite.

The Necessity of a Void

Einstein has given us a visible marker. All space within the universe's gravitational field is curved. Einstein's marker allows us to say that the curvature of the universe is its edge; beyond it stands the absolute void — uncurved, directionless, and infinite. If there were a limit to the absolute void it, too, would have an edge, curved or otherwise. We would then have to deny that it was an absolute void since an edge implies extension. Instead of what we thought was an absolute void, we'd be contemplating an envelope surrounding the universe which, unless it were absolutely separate from the universe, would have to be regarded as an extension of it. Without choice, we would have to follow the envelope to spatial infinity unless, once again, we came upon what we believed was an edge: a determination we could not make unless what lay beyond it were absolutely separate from the envelope. We could then say that here, here, here is the spatial limit of the universe. We could deny that the envelope was absolutely separate from what lay beyond it, which would oblige us to hurtle through a second envelope Since we could go on in this fashion forever, we appear to have two choices: one, we can imagine the universe as cocooned in an infinite series of envelopes which would deny that the universe is spatially finite; or, two, we can say that the curvature at the outermost reaches of the universe's gravitational field defines the spatial finitude of the universe: a determination we cannot make unless the universe is suspended within an absolute void: uncurved, directionless and infinite.

Only when something said to be spatially finite is found to be extended within a void can it with certainty be said to be spatially finite.

Protogenesis Within a Void

Our primary hypothesis needs to be modified then to admit the reality of the void. It would now read something like this: Fifteen billion years ago, a spatially-finite Protogenesis, extended within a void, converted all of itself into the pinpoint dot of the Big Bang. In the process, Protogenesis disappeared rendering it forever unknowable.

Despite such cosmic hijinks, we've learned a good deal about Protogenesis. We now know with certainty that it had to have been spatially finite; that it was extended within a void; that it had length, breadth, and depth; that it had a converter and a timer; that the delay between the state in which it was not converting and the state in which it did convert implies an interval which implies motion. We can't be certain that it had any other properties and attributes and we can never know the true nature of Protogenesis (if there were one); but we can be absolutely certain that whatever properties and attributes it might have possessed, they all had to be contained within the limits of Proto's length, breadth, depth.

Any and all properties and attributes of an entity must be inseparably fused with all other properties and attributes of the same entity, or we have to deny that those not inseparably fused belong to it.

Plucking a Point from a Line

It would be difficult to imagine a point along any span of length which does not belong to that length. We think that we can easily pluck a point out of that length but the evidence we would have for proof would be the broken original (therefore two separate lengths) plus the plucked point (therefore three separate lengths). Further, each of the three lengths also possesses breadth and depth. We think we can easily pluck a point of breadth, or depth, from any one or all three of the lengths, but we can't. No matter how small or large a point we think we can pluck, it always possesses length, breadth, depth. The illusion which imagination allows us of a two-dimensional reality is an error which conveniently allows us to forget that every point, line, wave and image has a backside.

We do the same thing with a line that we do with length, breadth, depth. We think that because we can pluck a point out of a line we have, therefore, broken the inseparable fusion of the line. But the inseparable fusion of a line is an illusion. The reality of the line is, as we know, an unfused aggregate of elementary particles. We're able to pluck a point out of a line because it's absolutely separate from each and every other point. If the line were an absolute fusion of elementary particles, we would never be able to separate them.

The notion that what is fused must be absolutely fused, therefore inseparable, and that what is separate must be absolutely separate, therefore unfusable, may be worth a brief digression.

The Impenetrability of Fused Entities

We've seen that for something to be separate it must be separated by an absolute void: a gap of empty space no matter how finely we draw it. To fuse two such entities requires that their edges be separable so as to be penetrable. To separate their edges, however, requires that a gap of empty space come between what we now can imagine to be segments of the edge. Unless the edge itself were separated from its core by a gap of empty space, we would be forced to report that it isn't just the edge which is separable but the entire entity. We would then be staring at an entity separated into a hundred or a thousand and more segments as though smashed by a hammer.

If our example were a pair of electrons (since electrons have thus far been unsmashable, they give the appearance of an absolutely fused entity), we would be left to imagine how two electrons so shattered might then be fused. As we began the task of putting the pieces together we would find that each piece has the same problem that the two whole electrons had to begin with: they cannot be fused unless the edges of the pieces about to be fused separate to permit the fusion. We could continue the exercise of separating the edges of the now tinier pieces, wondering, again, if their edges were separated from their cores. Deciding that they were, we would then smash them anew until we could smash them no more. We would have reached the absolute limit of their reducibility.

We would recognize this absolute limit when the most we could say about each piece is that it possesses length, breadth, and depth; that length, breadth, and depth are inseparable; and that this new infinitesimal mass describable as mere length, breadth, and depth is completely suspended within an absolute void. We would then need to say that it is absolutely separate from all other entities and that its properties (l, b, d) are absolutely fused, therefore inseparable. While the shape of this irreducible mass would seem to make no difference, the absolute fusion of length, breadth, and depth suggests that they be indistinguishable one from the other, which is another way of saying that they would be absolutely identical. One form in which length, breadth, and depth would be absolutely identical is a solid sphere. In this case, an impenetrable solid sphere.

Motion as the Fourth Dimension

The problem with a solid sphere is that elementary particles appear to be waves as well as solids, which raises the question of how a solid can also behave as a wave. Since in this instance our irreducible solid sphere of mere length, breadth, and depth is impenetrable and since exactly the same might be said of an electron, the clearest implication of the "wavy" action of an electron is that motion is an inherent property of the electron as well as our hypothetical little solid sphere. For motion to be a property of the electron or our sphere, however, requires that motion be absolutely fused with length, breadth, and depth, therefore inseparable from them and absolutely identical with them. We would then need to see motion as inseparable from length, breadth, and depth as depth is from length and breadth from depth. Motion, defined as the fundamental ability to move, would be the true fourth dimension.

The Alternatives to Motion as the Fourth Dimension

There are three apparent alternatives to the fusion of motion with length, breadth, and depth in the electron, in our little solid sphere, therefore in all elementary particles. One is the vision of existence as a spatially-infinite sea of undulating something whose vibrations we sometimes see as particles, sometimes as waves. Another is Newton's vision of motion as a "disembodied force," one lacking in length, depth and breadth which in various mysterious ways (collision, attraction, friction) acts on inert bodies causing them to move. The third apparent alternative is the modern view of "force particles" (gravitons, electrons and gluons) in which motion is indeed fused with length, breadth, and depth, a convenience denied to all other particles. The force particles then act in various ways (gravity, electromagnetism, strong and weak nuclear forces) on all other particles causing them to move.

Newton's Concept

In the vision of existence as a spatially-infinite undulating sea, it's easy to see why motion must be fused with length, breadth, depth; with every nook and cranny filled with the something of the sea, there is no alternative. The sea would have to be spatially finite to prove the alternative of motion as a disembodied force hovering around the edge of the sea constantly nudging it so that it might undulate. The problem here would be the same as in Newton's view of motion as a disembodied force. Where in this case it would hug the outside edge of the whole of existence, in Newton's case it hugged giant stellar bodies and particles. (Hugging is our word, not Newton's.)

The problem with hugging the whole of existence; or with hugging the sun, the moon, galaxies, protons or any other particle or body is that for motion as a disembodied force to hug them it has to cross the gap of absolutely empty space which allows us to distinguish them as particles, waves, bodies or the whole of the universe. Leave the frailest hint of a gap and we would have to deny that it had succeeded in hugging anything. On the other hand, if motion as a disembodied force were to cross the gap completely we would have to say that it was absolutely fused with the body, particle or universe since we could not discover a separation. We could imagine motion as a disembodied force advancing and retreating across the gap of empty space surrounding each mass — hitting and running, so to speak — which would then force us to consider its relationship to the void surrounding the whole of the universe or the whole of a particle.

Since motion as a disembodied force would lack length, breadth, and depth, it would lack an edge that might otherwise describe its position within the void. Lacking an edge because it lacks quantity (no quantum here), it would move freely in and around the infinite reaches of the void. Since by definition of a void it would meet no resistance, it is reasonable to assume that it would have "filled" the void infinitely long ago. As a result we would have to deny that what we think of as a void amounts to a void and we would need to resurrect Michelson and Morley to rerun their experiments for wind resistance as well as wind drift in the void.

Having denied that the void amounts to what we mean by a void, that is, the absolute absence of all possibility, but instead amounts to a spatially infinite presence of motion as a disembodied force, we have to deny that motion can hit and run. With no empty space, we would again have to see motion as a disembodied force hugging particles, planets and the universe. The new problem this creates is how motion as a disembodied force can hug a particle, body, or the undulating sea of the universe without fusing with the edges of each entity. With no way to retreat (it absolutely fills what we regard as the cold, dark void), the hugging is permanent. The permanent hugging of a particle, planet, or undulating sea amounts to permanent fusion. The permanent fusion of motion as a disembodied force with the length, breadth, depth of any entity denies its disembodiment and suggests a fusion as inseparable as the fusion of length with breadth and depth and depth and breadth with each other. The notion of motion as a disembodied force is not a tenable one, particularly since there is no experimental evidence of it. Motion always presents itself as fused with length, breadth, and depth.

The Alternative of Force Particles

Force particles present the same problem as disembodied motion when we try to understand how they interact with supposedly inert particles. A gluon, for example, binding the constituent elements of an atom's nucleus, must either attach itself to or penetrate the outer edge of a proton and a neutron to effect the bond. How does it accomplish this magical feat unless the segments of the edges of a proton are capable of separating to permit the penetration of the gluon? Easy enough, we think. The proton and the neutron simply open up their (surface) edges and the edge of the gluon enters. But unless we're willing to accept a meaningless entry, the force particle must touch, that is, fuse, with the now-open segments of the proton's or neutron's surface edge. Also easy, we think: it simply touches them. But the gluon cannot "touch" the edges of the proton's or neutron's open surface segments without crossing the gap of empty space that permits us to identify the gluon as a gluon and the edge of the proton or the neutron as its edge. To cross that gap, that is, to fuse, the gluon must penetrate the length, breadth, and depth of the proton's and neutron's segments and that it cannot do because the length, breadth, and depth of those segments are absolutely fused, therefore inseparable, therefore impenetrable.

Impenetrability Works Two Ways

The problem may be clearer if we switch perspectives and see the task of fusion from the gluon's viewpoint. How does the gluon impart its force to the open surface segments of the proton and neutron? How does it, the gluon, release, exchange, emit its "force" so as to bind the proton or neutron to it? The gluon is theoretically a fundamental force particle in the same sense as an electron or a quark are elementary particles which means that the gluon has no constituent parts, is therefore irreducible, therefore impenetrable.

But impenetrability works two ways. Emission is as impossible as admission, particularly since the gluon has no constituent elements to emit. The *whole* of the gluon must therefore insert itself into the length, breadth, and depth of the fundamental *fiber* of the proton or neutron and that it cannot do because the fundamental fiber of the proton or neutron is the very inseparability of its length, breadth, and depth. Why can't the gluon simply attach itself and cling to the proton's or neutron's edge? Because attaching and clinging are more permanent stages than touching. To cross that gap of empty space between the gluon and the edge of a proton or neutron implies fusion of the four dimensions of the gluon with the three dimensions of the proton or neutron. Somehow, we've got in the habit of thinking that motion may be disembodied whenever it is convenient for us to require it. Certainly in a force particle, motion cannot be separated from its length, breadth, and depth anymore than the length, breadth, and depth of a proton or neutron may be separated to permit the commingling of the gluon's inseparable dimensions.

Arguing that the proton may be an illusion doesn't help. Breaking the proton up into its "parts" only makes matters worse. When we finally arrive at the quark we have what is admittedly an elementary particle; therefore a particle without constituent parts; therefore a particle which is irreducible, impenetrable and non-emitive. How does a gluon interact with a quark? How does it penetrate the impenetrable edge core of an elementary particle like the quark? It can't. There is no way, logically anyway, for a quark or any other elementary particle to *receive* or *admit* the force of a gluon *even* if the force of a gluon could be disembodied. No one suggests that it can. The suggestion is that the strong force is absolutely contained in a *particle* so that it is the gluon particle which would have to be received or admitted by a proton, neutron, meson, muon, pion, or quark. And that cannot be done.

The only reasonable explanation of force is to discard the notion of it either as a disembodied entity or a special particle and to see it, instead, as the result of the fundamental fusion of motion, length, breadth, and depth. The fusion of motion, length, breadth, and depth allows us then to see movement as *self-generated.* It also allows us to see why elementary particles can — indeed must — behave like waves. It also accounts for the infinite variety of shapes and forms elementary particles seem to assume.

Structuring the Folly of Protogenesis

The rule regarding separation and fusion is another way of demonstrating the axiom — yet to be proved — that something can't become something else. The rule is particularly useful in the attempt to structure a Protogenesis. We saw that for Protogenesis to have converted itself fifteen billion years ago, it would have needed a property which, for want of a better name, we called a conversion factor. The conversion factor is mysterious and totally strange to us. The closest analogue would be God's ability to transubstantiate His being into the being of matter. But that ability must also be innate to Protogenesis, otherwise we would have to deny that the pinpoint dot of the Big Bang ever emerged from something other than itself.

Protogenesis would have also needed what, again for want of a better name, we called a timing factor. Protogenesis, sitting around for a period infinitely backward in time, suddenly, fifteen billion years ago, converted itself into a pinpoint dot which suddenly exploded. What else would we call such a factor? *Something* delayed the conversion of Protogenesis for an infinity of time. Or something required an infinity of backward time to trigger the conversion. Again, the closest analogue in the divine theory of creation is God's Will. God simply decided that He wanted to create a universe and He willed it. In the non-divine version, the implication that a timing factor would have a tick-tock factor measuring the events leading to the conversion of Protogenesis is also inescapable. It might have been a tick-tock factor totally strange to us but something had to control the interval between when Protogenesis remained Protogenesis and when the hypothetical conversion took place.

Fusing the Conversion and the Timing Factors

Like it or not, Protogenesis amounts to a dreadful duality. On one hand we have something which remained at rest for an eternity and which suddenly converted itself into a pinpoint dot. There is a sneaky implication here that the timing factor itself might have suddenly manifested itself — separately — out of nothing, which burdens us with the necessity of a second creation out of nothing, or two Protogeneses again. It would be wiser to examine first whether or not the duality itself holds up to reason.

For example: Where would we position the conversion factor and the timing factor so that they were not inseparably fused with the length, breadth, and depth of Protogenesis? If they were inseparably fused they would amount to the same factors. The meanings attached to each, however, deny that they are the same factor. The conversion factor is seen as a static entity "at rest" for an infinity of backward time; the timing factor is seen as a tick-tock factor accumulating or delaying for an eternity the conversion of Protogenesis. The conversion factor has the characteristic of mere length, breadth, and depth. The timing factor has the characteristic of length, breadth, depth, and motion — fused. The conversion factor lacks the fundamental ability to move; the timing factor does not.

If they were inseparably fused the conversion factor would also possess motion, and the particular motion of a tick-tock factor. On the other hand, if they were absolutely separate, the timing factor would be faced with the task of how to penetrate the length, breadth, and depth of the conversion factor without fusing with it. We have seen why motion either as a disembodied force or as a fused property of a force particle cannot touch or penetrate a so-called inert entity. It cannot penetrate simply because the length, breadth, and depth of any entity are the very fiber of an entity and they are inseparable.

Obviously, something is wrong. For Protogenesis to have converted itself requires that the conversion factor and the timing factor amount to a single fused factor. We would then have a Protogenesis in which the ability to convert itself into a pinpoint dot had been absolutely fused with the factor which could initiate such a conversion an infinite number of times but somehow did not. Worse than that, the conversion should have occurred the very split instant *Protogenesis* was created. When might that have been? Well, one supposes, when infinity began. Mired in a welter of self-contradictions, theoretical physics asks us to accept the "preposterous." After all, as one cosmologist put it, "it only had to happen once." The sad fact, the logical fact, is that it never could have happened at all.

Why Something Cannot Become Something Else

To have converted into a pinpoint dot, or anything else, Protogenesis would have had to convert itself into its *absolute opposite*, that is, something which did *not* possess length, breadth, depth and motion. The complete and absolute opposite of extension (length, breadth, depth, motion) amounts to non-extension or what we mean by a void: nothing, empty space, the absolute absence of all possibility. Why would Protogenesis have had to convert into its complete and absolute opposite? Why couldn't it have converted into its *almost opposite*? Because the *almost opposite* of something said to be extended amounts to something which *almost* possesses length, breadth, and depth (always with motion).

While virtual particles and ghost particles seem to fit the bill, they don't. Ghost and virtual particles possess length, breadth, depth and motion. To have converted at all, Protogenesis would have to have converted itself into nothing, that is, caused its own disappearance, which would deny the theory. The theory says that it converted into a pinpoint dot of such enormous energy that it had to explode. *Could* Protogenesis have caused its own disappearance? Not by a mile. Why not? Because it was spatially finite. To have been spatially finite it had to have been completely surrounded by the void so that, as a whole, it was already as deeply *into* nothing as it could possibly be. We could imagine its break up into trillions of tiny pieces but, no matter how small, each would have had to have been a whole Protogenesis and each would have had to have been completely surrounded by a void, or we would have to deny that they were absolutely separate and we would, therefore, have to deny the break up. Completely surrounded by the void, each minuscule whole of Protogenesis (as with the unbroken whole of Protogenesis) would have been as deeply into the void as it could possibly be. Why couldn't the void do something to it? Because nothing cannot amount to what we mean by nothing, that is, the absolute absence of all possibility, and still contain the possibility of absorbing, hiding, warping, or otherwise acting on Protogenesis, our universe, or anything else.

Protogenesis could not have converted itself into the pinpoint dot of the Big Bang simply because it could not have changed. To have changed would have required that it become nothing. Change of any kind amounts to an impossibility. Something cannot become something else simply because the only possible way something might change is to become nothing. Since reality includes a void, everything is already as deeply immersed in the void as it can possibly be.

The Universe Always Was and Always Will Be

There may have been a Big Bang and it may have happened within a pinpoint dot, but one logical necessity seems fairly certain: the pinpoint dot did not suddenly or gradually appear, emerge, or spring out of nothing, or out of something else. The universe always did and always will exist. If it is expanding, then it will also absolutely contract. The alternative is to believe in a one-time creation (the Big Bang as a singular event, a singularity, the current theory) or otherwise be stuck with the implication that the universe has been expanding from infinite backward time but that infinite backward time hasn't quite been long enough to accomplish what is now about to be accomplished: the imminent end of the universe's expansion twenty, thirty, one hundred, or one thousand billion years from now. Expansion implies movement from a center. The only way a *timeless* universe could have moved from a center is to have formed it by contracting. And the only way it could have contracted is if it had expanded.

The evidence of galaxies moving outward at speeds approaching the speed of light combined with the evidence of ancient radiation clearly implies a distant explosion some fifteen or so billion years ago. What it does not imply is the singularity of that explosion. If the bang happened, it has to have happened as the contraction of the universe, not as the emergence of the stuff of the universe as a pinpoint dot after roiling and boiling for eons backward in time. Ultimately, the compression of the stuff of the universe prior to the one-time big bang isn't any different than the repeated contractions of the universe culminating in infinitely repeated and repeatable Big Bangs.

Chapter 2. Walking through Infinity

We know now why the universe could not have had a beginning. The principle of continuity states that something cannot come out of nothing. To have had a beginning, the universe would have had to arise from something other than nothing and other than itself, for to have arisen out of itself would implicitly deny that it had a beginning. So to have had a beginning the universe would have had to arise from something else. As we have seen, the principle of continuity denies that something can become something else for to become something else the universe would have had to become its absolute and complete opposite, which amounts to nothing.

The first law of thermodynamics also states that energy can neither be created nor lost. So unless we are willing to accept that the universe repeatedly expands and contracts, and does so infinitely in time — or unless we're willing to accept that there is no expansion and contraction, therefore no Big Bang — we have no choice but to accept a tamer, calmer universe than we've grown used to in the past thirty or forty years.

Einstein reportedly believed that the pinpoint dot created its own space (nothing) as it emerged and exploded. Sort of like a shadow, but an empty shadow, which keeps being pushed out with the expansion of the universe; an empty envelope, so to speak, which would roughly depict curved space. In effect, he believed that prior to the Big Bang singularity there was nothing. No empty space. No void. So the void had to be created.

But for a void to be created requires, in this instance, that it be created out of nothing. Relativists might protest that there was no nothing — that the Big Bang created nothing. What they would be arguing is that absolute non-existence does not amount to nothing. If not nothing, what would it amount to? A vacuum? Yes, vacuum genesis. Dance as we might around the May pole, a vacuum still amounts to nothing and nothing cannot create nothing.

Even if it could, even if it hugged the dot like an empty shadow or an empty envelope, we would have to infer an infinity of empty space which would allow the exploding pinpoint dot all the elbow room it would need to expand.

Emergence Implies a Context within which to Emerge

If we allowed the creation of an infinite void simultaneously with the creation of the pinpoint dot, the nagging question would be within what was it created — which is perhaps the riddle Einstein was trying to solve. It was obvious that the pinpoint dot had to emerge within some context. Merely to utter the word "emerge" implies a context in which the emerging would take place (as well as a context from which to emerge). For some reason not apparent to us who do not understand either the specific or the general theories of relativity, Einstein could not accept the presence of an infinite void within which the pinpoint dot could emerge. The absolute presence of an infinite void must seem offensive to some physicists; one rarely comes across a reference to it.

An Infinitely Dormant Void?

An infinite but dormant void would be especially offensive when one is proposing the singular creation of the universe. For example: Why is the stuff of the universe being created for the first time when standing there waiting for the pinpoint dot to emerge for the first time is the infinity of the void? Empty space. Waiting. Waiting. Waiting. Forever. Waiting for the extraordinary moment when a pinpoint dot of something will explode within its confines. It's sort of like asking why God took so long to create the world. What was he doing for the eternity before creation? Actually, it makes more sense to ask what God was doing for eternity — eons and eons of time — than to posit the empty vastness of an infinite void waiting for the Big Bang singularity. God had heaven and angels to occupy him until he had the spectacular idea of creating a universe.

The Problem Is the Presumption of a Beginning

One can hardly blame physicists for rejecting the possibility of a dormant infinity of empty space waiting for the great moment. The problem of course is the presumption that the universe had to have had a beginning which, as we have seen, was dictated by the evidence of stars exploding outward in space and signs of ancient radiation flashing in the darkness. But the physical evidence of red shift in the outward flying stars and ancient radiation is more easily resolved by positing the repeated expansion and contraction of the universe within an absolute void. There must be a good reason, or reasons, why physicists reject the repeated expansion and contraction theory.

One of the problems that infinite expansion and contraction would answer is: The edge. As we saw in the preceding chapter, things said to be finite must have an edge in order to say that they end here, here, and here or there, there, and there. We saw that, even to be imagined as a dot, pinpoint or not, the Big Bang singularity needs an edge. We're free to describe the dot any way we please, but we cannot escape the fact that to be called a dot it needs an edge.

Perceiving an edge requires a frame of reference, a context, which would allow us to say that the dot ends here, here, and here. Otherwise we cannot assume that what supposedly emerged was a dot. An ellipse, a square, anything would still require an edge. An edge is what allows us to define things. One of my critics reminded me of Einstein's proof that finite things do not require an edge. Einstein reportedly said that if a pair of two-dimensional people were laid flat on the surface of the earth, they could not perceive an edge. I replied that I do not know what two-dimensional people are but I would wager that my friend could not depict them without an edge. If what emerged, then exploded in the Big Bang, was a dot then defining it as a dot requires an edge and an edge requires a void within which to frame the dot so that we might say that it ends here, here, and here in the form of a circle. Such a void would also have to be infinite for the same reasons. So we're back to infinite space.

The Void Must Be Infinite

We would now have to imagine a void reaching infinitely in every direction. We would also have to imagine something within the void with the potential of becoming a pinpoint dot which itself would contain the potential of mushrooming into the billions and billions of galaxies of the universe. If the void contained anything that had the potential of becoming our universe it would not, of course, amount to a void. What we mean by a void amounts to the absence of all possibility. It could not, therefore, contain the potential to become our universe or anything else. Nothing cannot become something. But it could, for example, be filled with energy.

By energy, I mean what Nick Herbert called the stuff of the universe (*Quantum Reality: Beyond The New Physics*). The finest dust you can imagine. (Herbert didn't call it "dust," he called it "stuff.") The dust which fills the spaces between the electrons in the outer ring of the atom. The dust which is released in an atomic explosion and the controlled chain reaction of a nuclear engine or power station.

We could imagine this infinitely boundless condition to be the primordial condition of the universe; the end phase or the start phase of expansion; the alpha and the omega of existence. In its start phase, it might well be regarded as one of the possible Protogeneses. But as we saw in the preceding chapter, Protogenesis needs a clock, what we called a tick-tock factor, and a converting mechanism; a clock to dictate when to end the expansion phase and when to start the phase of expansion. Physicists seem to believe that when the ultimate particles have cooled off they start to collide and the collision results in the formation of atoms and from there the formation of complex aggregates of atoms like suns, galaxies, etc. (See Stephen Hawking's *A Brief His-*

tory of Time, pp. 108 ff). Two questions. One: What causes the cooling down of the ultimate particles if there is no void — if everything is filled with energy? Two: How can a particle of energy lose heat when it is the ultimate particle and is presumed, therefore, to be impenetrable? Assuming that the ultimate particle of energy is not impenetrable and that the answer is radiation, how does radiation lose heat and why isn't radiation the irreducible condition of existence? Assuming that radiation is the ultimate condition of existence and assuming that it can lose heat and, losing it, begin its coalescence into the atom, we can say that in an infinitely extended universe, chock full of radiation, the tick-tock factor and the converting trigger are one and the same: temperature.

Is the Void Chock Full of Energy?

So we have the possibility of a universe which in its alpha-omega condition is filled with radiation and contains the four dimensions of length, breadth, depth and motion (l, b, d, m). But being chock full of energy, as we supposed, the universe would be static, would it not? Not if energy possessed the dimension of motion. Lord, if anything might be said to possess motion, it would certainly be the ultimate particle of energy. But where would the ultimate particles of energy move if they filled the universe to choking? No matter, the particles would soon begin to coalesce, but in coalescing, would they not leave a void? Indeed, they would. A void would also be required if we were to distinguish the various forms of coalescence. How could we distinguish what was an electron, what a proton, what an orbit, what a nucleus, without an edge to say that this one ends here, here, and here; that one there, there, and there? And how would we distinguish each edge without an empty space — no matter how thin — separating them? We couldn't. Everything other than infinity or nothing must have an edge. You may think that walking into the sunset you could see no edge, but most of us take the horizon to be an edge.

The horizon is also what taught us to believe that the planet was round. We also now have full color photographs of the planet taken from outer space which reveal that the edge of our planet is its circumference. Lo and behold, the earth is round. So are the sun, the moon and our sister planets in the solar system. Indeed, the universe itself appears to be a gigantic bubble expanding equidistantly in all directions from a center. (Sir Arthur Eddington once described the universe as an expanding pebbled balloon, the pebbles representing the billions and billions of galaxies with the unfilled spaces representing the void.)

Filling in the Spaces

What surrounds the expanding balloon? In its immediate environs, space curved by gravitation which implies a particle field. And beyond that, well, the infinite void without which it would be impossible to imagine or describe the phenomenon which Eddington so felicitously likened to an expanding balloon. So there we are again, faced with an infinite void. Should we now add an infinite void to the alpha-omega of existence? But can we accurately describe infinite nothingness as a condition of existence? Not so anyone would accept the description. To be accurate we would have to describe the void as the condition of non-existence or, drawing the definition more finely, the condition which permits us to individualize existence — to say this object is a sun, that one a moon. Imagine the mess we'd find ourselves in if objects merged into one another; the universe would be an enormous swamp. We, on our planet, are separated by air but it wouldn't be hard to imagine how cumbersome our lives would be if we merged into one another like Siamese twins, joined not just at the hip but everywhere on our bodies; we'd probably suffocate, or grow into monstrous creatures.

An infinite void chock full of energy would make the compression of matter hypothesized for the pinpoint dot seem plausible. The problem is that being infinitely extended it would still be compressing energy into the pinpoint dot. Or maybe the dot wasn't a pinpoint. Maybe energy merely contracts until it reaches the point of combustion. Why does it have to be a pinpoint dot? And why does all the energy in existence have to contract in order to have a singularity? It couldn't, if the void were infinite and jammed with energy particles. Why not? Because if it were infinite, it would still be at it. But if there were a point of intolerance in the process of contraction, a point where it would simply explode, there would then be an infinite amount of energy left to contract and then explode so that existence would seem to be a fireworks display with universes popping all over the place.

Like a Flotilla of Eddington Balloons Popping All Over the Place

Wouldn't it be crazy if what we call the observable universe turned out to be just that? — Not simply what we could see of the universe at the moment (implying bigger and better-placed telescopes) but a small corner of an infinite void chock full of energy where universes smaller and larger than ours are contracting and expanding all the time. It would be even crazier if instead of a one-time Big Bang a lot of little bangs were happening all the time, right here under our very noses.

If the universe has been repeatedly expanding and contracting within an infinite void, the nagging question would be, Which came first: the expansion or the contraction? Is it a reasonable question? It would appear so unless we assume that the fundamental condition of the universe is a steady but roiling mass reduced to ultimate particles of energy which every trillion years or so begins to coalesce; in which case the answer would be obvious: contraction. Of course, we'll never know. All we can be sure of is that nothing cannot become something. No, we can also be sure that nothing cannot become nothing. We could then walk through infinity, walk, walk, walk, walk, slowly shedding our primitive night and day notions about beginnings and endings, while we grew comfortable with the idea of infinity.

CHAPTER 3. FALLING TREES

Since something cannot become nothing, since nothing cannot become something, and since something cannot become something else, the obvious conclusion to which the principle of continuity leads us is that the universe amounts to one kind of being. The principle supports what we call "the material universe" and it would appear also to support a single existence of spirituality, or what some of the most distinguished philosophers of history have referred to as idealism.

Absolute Idealism

Plato (428–348 BC), to begin with the most distinguished pundit, believed that while stones, trees and oceans seem to be real, which is to say they enjoy an independent existence, they are no more than "shadows" or appearances, what philosophers call phenomena. Ideas, not things, are the true reality and ideas exist in the mind, which is an attribute of the soul. Ultimately, therefore, ideas exist in the mind of God. In effect, there is no world of objects; there is only a world of ideas.

Subjective Idealism

A less monistic form of idealism, called subjective idealism, was devised by a philosopher turned cleric named George Berkeley

(1685–1753), often referred to as Bishop Berkeley, who lived two thousand years after Plato. Berkeley is best remembered as the promulgator of the riddle which asks: If a tree falls in the forest with no witnesses to the sight and sound of the fall, who is to say that it fell at all? No one. In short, argued Berkeley, the existence of something that we believe to be out there — an object with an independent life of its own — depends wholly on a perceiver. We cannot be certain that there are trees in a forest, let alone falling trees, until and unless someone perceives them.

God as the Ever-Present Perceiver

To Berkeley, the ever-present perceiver is God. Can we say that the Atlantic Ocean is kicking up one hundred and thirty foot waves this morning? How can we find out whether or not a storm is raging in the middle of the Atlantic or, indeed, whether or not the middle of the Atlantic is still there? We rely on weather reports, satellite photographs, the reliable perception of a perceiver. However, the certainty that the middle of the Atlantic is still there when satellite cameras are not focused on it depends on God as the ever-constant perceiver.

Bishop Berkeley's argument was a reaction to the explosive rise of the scientific age. Copernicus contended that the sun did not revolve around the earth; it was the earth which revolved around the sun. Galileo and Newton proved that all objects obeyed discernible laws of motion. Cutting to what he perceived to be the bone of scientific method, Berkeley set out to prove that the attempt to substitute scientific method for Revelation was an illusion. For what science assured us was an independent world of objects was no more and no less than an illusory world of perceptions dependent on God, the ultimate perceiver, for their everyday, minute by minute, existence.

The Presumption of a Self

The problem with both absolute idealism and subjective idealism is the presumption that what we regard as a self is essentially a soul/mind. But if we were to make no assumptions about what consciousness or awareness might be, we would find that we cannot empty the container we call a self, or perceiver, of what we call objects or perceptions. Stand still for a moment and try to empty yourself of the objects you see, hear, smell, taste and touch. We can close our eyes and the moment of experience is filled with darkness but we can still smell a steak cooking or jasmine blooming, and we can hear the wind rustling through the trees. Try though we might to empty the moment of objects, we cannot. We could stand defiantly for a hundred years staring into the woods or at a crowd of people in a stadium and we would never isolate what we think of as an inner self, let alone a soul/mind. All that experience gives us is a world of objects with no affirmative or contrary evidence that they exist independently Same with our sense of a self. It is easy to think of it as a mind, an attribute of the soul we have been led to believe is the animating force of life, but we can find no evidence of it.

A Self Is Not Revealed in Experience

The only evidence the moment of experience gives us is the content of the container, the world of objects. We could call them ideas but then we would be presuming that there is a self and that it is a soul/mind. The evidence of experience is that a self is not revealed in experience and therefore that it does not exist. What exists are what we call objects. We might argue that calling the world of objects rather than ideas presumes that what we call our sense of an inner self is a brain. But there is more to the sense of a self than a brain. All we need to do is to stick a needle in a finger to discover that a finger belongs as much to a sense of self as a brain. It is, after all, a rather recent discovery that the brain is our thinking, reasoning, remembering organ. All that experience gives us is a world of objects which exists *a priori* (before we start to wonder about it) and which cannot be obliterated, therefore a world which exists independently. We could easily obliterate it or have it obliterated for us by an infusion of anesthesia or we could commit suicide when we are driven to despair by the irreducibility of the world of objects, but then we would be missing a player for our soccer team or a fourth for bridge. In any event, we should have no doubt that when a tree falls in the forest it falls without the aid of a perceiver, divine or terrestrial.

Meditation Does Not Reveal a Self

Still, Buddhists meditate with the very purpose of emptying the mind of objects. They claim to achieve this blissful state when the mind is aware of nothing. Perhaps a hum. Or the sound of one hand clapping. But is this what we mean by a perceiving mind? Where are the sharp tools of the intellect? Where are the naked attributes of mind? Is the stripped-down mind aware of God?

God as a Mathematician

Plato believed that mathematics was one of the proofs of God's existence. Philosophers have argued that the world of objects does not give us the foundations of mathematics. Nowhere in the world of objects will we find a pure circle; a pure circle is an invention of the mind. So are numbers. As are the many diagrams of mathematics (cones, triangles, etc.). Mathematics is innate to mindness, God's and ours, so argued Plato. But without the ability to isolate a self which we could fairly describe as a soul/mind, we have little choice but to bend to the evidence that reality consists of a world of independent objects and that mathematics, too, is likely derived from the processes in which the atom engages, an unconscious habit no different from the process of logic. In effect, the innateness of logic and mathematics can just as easily be ascribed to the processes of nature, the world of objects, as it can to the mysteries of the soul.

Buddha the Materialist

In fairness to Buddhists, we need to emphasize that Buddhist meditation does not have as its purpose the experimental discovery of a mind (Buddha did not believe in mind/soul/spirit; he was a thoroughgoing materialist) but rather the exorcism of objective reality, which Buddha believed to be the source of all sadness. But it is also the source of all happiness. If happiness is ephemeral and must ultimately succumb to death, then we must conquer death, not deny life. Buddha has been described by scholars as a "world-weary" Hindu tired of being reincarnated and longing for the long dark repose of Nirvana. He has also never been duly honored as the first materialist. He believed that existence consisted of "heaps of atoms" one hundred years before Democritus.

Plato's Old Souls

The belief that we are possessed of a perceiving self which is essentially a mind or a soul or spirit is therefore a presumption based on the nature of unexamined reality. Awesome as he was, Plato had little patience with the analysis of experience or what we would call the examination of reality. Plato scoffed at the possibility that knowledge was to be gained from observation of reality. Instead he believed that the soul was constantly, and eternally, recycled so that our souls were "old souls" full of knowledge. As a result, knowledge was recollection of "old knowledge" and the life of the mind/soul the only reliable source of what was real and what was not.

The embarrassing implication of knowledge as the recollection of an "old soul" is its failure to reveal knowledge which we have desperately longed for and which we have gained, and continue to gain, through trial and error in the world of objects (scientific method). Where was the old soul at the time of the bubonic plague; where is it now as we search for a cure for cancer? Why did it take thousands of years to discover bacteria and viruses? Why so long before we could see that the double helix was the secret of life? Why did the old soul mislead Aristotle into believing that human reproduction was caused by a homunculus (a little man) in the semen?

The Probable Origin of Belief in Souls

The belief that we possess souls and that souls are the motivating force of life was probably a mistaken perception by our primitive ancestors about the cause of death, which they variously described as escape of the soul from the body, measured as loss of breath. In some places, holding a mirror up to the mouth and nose of a putative corpse is still the measure of whether or not a person is dead. Some of our primitive ancestors likened the departure of the soul to the flight of a bird from the body (see James Frazier's *The Golden Bough* for other examples). In Genesis, God is revealed as having imparted his own breath into the dusty figure of Adam to give it life. God would not have revealed anything so inane.

God Would Have Known that Something Cannot Become Something Else

God would have known that his being, imparted as breath, could not have been the animating force of life. Not because men and other animals were undeserving of even the smallest spark of God's being but rather because God would have known that something cannot become something else. To become the animating force which is life, God's breath would have had to fuse with the billions of cells in every body and then fuse with the thousands of genes in every cell, in effect creating a perfect clone of himself. God would not have done that. He would have known beforehand that to fuse with a cell his breath would have to bridge the gap between a cell and God's breath. In effect God's breath would become the cell, then become the gene, consuming everything in the cell as it permeated the body, thereby filling every possible space and thereby obliterating the image he had just made of dust, and turning it into God. God would have known that what appears as a process of becoming is an illusory process, which is to say an unexamined process.

The Interaction of Soul and Cell

Except in absolute idealism, we do not assume that God's breath and the cell are fused. We think that our souls enjoy a parallel existence with cells and brains, but they don't. To interact with a cell rather than fuse with it, God's breath must again bridge the gap between cell and soul and then penetrate every last gene, pausing at each level of its generation to consider how it might leap through the gap between itself and the last nucleotide, and in pausing see that it cannot penetrate a nucleotide or a cell. For in penetrating to interact with cells and nucleotides, God's breath must penetrate every last fragment of every last nucleotide or leave billions of fragments untouched. Then it must retreat from every last fragment of the cell (in and out, in and out) or else remain permanently in absolute occupation of every last fragment, thus fusing anyway and thereby denying the co-existence of body and soul and transforming the dusty figure of Adam and his progeny into a perfect clone of God himself. Where we insist that cell and soul might co-exist without penetration and without fusion, we think that the soul might impart its animating force with staccato bursts of divine energy. What we would have created then is a monster that would be dead one second and alive the next.

An alternating current of energy is not, however, the way we think soul and body interact. Like the absolute idealists, we assume that every last speck of our beings is filled with God's spirit breathed unto Adam and, presumably, breathed into each body as it is conceived. We believe that we are filled with God's being, filled to the ultimate particle in the least atom in our bodies with the indestructible being of God; therefore the guarantee that we do not die when our bodies die. "The Kingdom of God is within you," said Jesus. The question we forget to ask is, What could possibly cause the death of a cell fused with the being of God? Easy, we think. God simply defuses cell and soul, draws the soul — His being — into heaven and leaves the once-sanctified cell to rot in the grave. Worse, where we believe that body and soul may be locked together in the grave, it is God's being that we allow to decompose in the box; or, not decomposing, remain the immortal being of God imprisoned forever in a mud hole.

No Immortality in the Old Testament

The irony is that the Old Testament, or the first five books of it, was written by a high priest of Judea who did not believe in the immortality of the soul. He wrote the Torah/Pentateuch after the destruction of Solomon's Temple in 586 BC to renew his peoples' faith in God; to reassure them that God still loved them and held them in a special place in his heart. What he neglected to deal with was the immortality of God's breath. The belief in the immortality of the soul was so abhorrent — so Egyptian — to this ancient priest that he had God post his (God's) favorite angels around the Tree of Life to keep Adam and Eve from eating of the Tree "lest they (Adam and Eve) become gods like us." Adam and Eve would have had to eat from the Tree of Life in order to possess immortal souls. But God's breath, the breath with which he animated

Adam's dusty figure, had of necessity to be immortal simply because it was God's breath. We think that God can separate his being from his breath because he is God, but God would never have suggested such an incongruity. God would know that his breath was inseparable from his being. Was it God's breath or was it not? If it was, then it must be suffused with God's being.

Berkeley's Folly

Berkeley's hope that he might have saved Revelation and therefore religion from the ascendant tide of science was futile except in the salons of bishops and friendly philosophers engaged in the dilettante's sport of theological discourse. The rest of us are left with a monistic existence in which we have come to accept that, eons ago, the lowly atom found that it could reproduce itself and in the process also discovered that it could think — reason, argue, imagine, speculate, deduce, induce and create: gods, devils, tools, clothes, paintings, pyramids, chariots, automobiles and, God bless them, children.

Chapter 4. The Improbable Being of God

The ancient Greeks, who otherwise produced giants in philosophy, felt it necessary when constructing a religion to posit Cronus and Rhea as parents for Zeus and Hera. They did so probably because they, like us, believed that all things must have a beginning. Why they forgot to apply the principle to Cronus and Rhea remains a mystery.

The ancient Judeans did not make the same mistake. In fabricating Jehovah, the writer of Genesis allowed us to believe that Jehovah had no beginning. He allowed us to assume that Jehovah was what later thinkers would call a "perfect being," which is religious jargon for Jehovah's omniscience, omnipotence and, of course, his infinitude.

Jehovah as God

The writer of Genesis did not explicitly infer these attributes to Jehovah. He did not have to. He assumed, correctly, that his congregation would infer, as the rest of us have, that Jehovah had to have been all the things he would have had to be to do all the things he would be said to have done. He was the creator of the earth and all of the wonders, animate and inanimate, therein. And he was the creator, above all, of humankind. For make no mistake, the Good Book is about us and all the things which distinguish us from all other living creatures:

two-legged animals who speak, think, remember, and have opposable thumbs which allow us to be weapons makers, cities builders, writers, mathematicians, clothes makers and wearers, house builders, inventors of wheels, discoverers of fire, builders of boats small and large, inventors of sails and oars, and last but not least, the discoverers of agriculture. Truth be told, our accomplishments were/are the living proof that we were/are extraordinary creatures necessarily come into being by some unique force, not found in nature, therefore supernatural; we were/are the very proof that God exists and why we are semi-divine in nature.

Jehovah's Continuous Being

By avoiding a beginning for Jehovah the writer of Genesis automatically skirted the absurdity of having Jehovah come into being out of nothing, or nothingness, as Sartre would have it. Coming into being out of nothingness was nonsense we were to be spared until the middle of the twentieth century when it was re-discovered by the inventors of the Big Bang. It is nonsense for it denies what we propose as the meaning of nothing: the total and absolute absence of any and every potential, therefore the potential to become something. Some might argue that the very ability to refer to nothingness also presupposes its creation, let us say, out of pre-nothingness, which is to say "the potential to become nothingness." In the opening to his Gospel John had some such pontifical nonsense in mind when he declared that, "In the beginning was the Word and the Word was God." What would John have had us believe — that the potential to become something existed as language, which would suppose a sensible language, therefore a mind, therefore a thinking existence which would become God? Or was it already God? John appears to have been disabled by a fit of poetic writing which inevitably muddles meaning.

A Crazy Idea

Einstein was the only serious thinker to propose that nothingness might have been created. Theoretical physicists love to propose crazy ideas as radical platforms which provide fresh jumping off points for more craziness. What else could the idea that nothingness was creatable be? Einstein seemed to be struck by the need for empty space for without it, it is impossible to distinguish the "edge" of something; say, an asteroid or a planet or, as in this instance, the edge of the pinpoint dot of the Big Bang. The truth is that space, or nothingness, had to have preexisted to allow the very recognition of a pinpoint dot for without an empty frame of reference it would have been, and would be, impossible to distinguish the edge of a pinpoint dot either in external reality or on the blackboards of our minds.

No Beginnings

The problem comes with the compulsion to demand beginnings because everything around us seems to come into being like buds on a fruit tree or fetuses in human beings and other mammals where they did not exist before. It is a habit of mind it would behoove us to break, lest we continue to slide on the slippery slopes of handy thinking. A beginning always implies a coming into being out of nothingness and we know that nothing cannot become something. Always. Otherwise, it comes into being out of something else and is therefore not a beginning.

A Divine Existence?

God is subject to the same rules, for if he exists then it is a rule he created. An omniscient God would know that God cannot point a finger out into the void (nothingness) and ignite the darkness ("let there be light"). The void contained the potential to become light, in which case it could not be said to have been nothingness. Or, if the void did not contain the potential to become light, then the finger of God contained such a potential. Light, therefore, had to have been created out of the divine substance of God. The same would be true of everything Genesis says God created: all of it, earth, sun, moon, stars, trees, water, people, animals, insects, the entire content of Genesis' six days of Godly labor; all of it divine in every iota: for not one speck of divinity may be said to be divine without being suffused with the being of God. God would necessarily know this and would therefore know — if he existed at all — that the story of creation fantasized in Genesis was false and not the true account of our imagined creation. If there is a God, then it is the height of irreverence to claim that he created the universe according to Genesis.

Absolute Idealism

Philosophers call the depiction of reality as relentlessly divine "absolute idealism," partly to distinguish it from "subjective idealism" and partly to convey the notion that everything we find rock hard and objective in nature is no more and no less than an idea in the mind of God. Truth be told, however, the greater logic appears to be not that mountains and trees are merely ideas in the mind of God but rather that God is a tree and a mountain and clouds and dirt and sand. (This echoes Spinoza's ingenious version of pantheism; itself an echo of the ancient Sumerian belief that their gods occupied special places in nature. Tammuz, for example, dwelt in a tree.)

You therefore have a choice: if you believe that God created the earth and all that it contains, as well as all that surrounds us, you may elect to be an absolute idealist like Plato or a pantheist like Spinoza. Either way, it will be as close as you will come to a rational belief in God. You will have to deal with the necessity of accepting things like feces and urine as divine; or worse, be confronted with the evidence that what we call good and evil are equally divine in nature, with evil usually triumphant.

To Be Divine Is To Be Immortal

Infinitely worse, and morbidly depressing, would be the evidence that though we are suffused with the being of God in every speck of our existence we remain mortal, and that is the first logical evidence we have that absolute idealism is not a viable explanation of God. Pantheism isn't much better. Pantheism is token religiosity merely duplicating what we know and can know of old-fashioned materialism by scientific method.

One cannot be divine in nature and be mortal. The Biblical notion that God fashioned us of dust and then breathed his breath into us to give us life is a heady fancy but false for the dust itself had to have been created by God and had therefore to have been divine in nature, filled with God's being, because the void was said to have been a void; sheer nothingness; therefore lacking the potential to become dust devoid of divinity.

All we have to do is to follow Michelangelo's finger of God reaching into the darkness and try to separate from his finger the attribute which gives it presence, its thereness: whatever we attempt to peel away is filled with God's being. We think that God being God can merely wish something to happen and, presto magic, it happens; an abracadabra God. What God cannot do is separate the attribute which gives him life, presence, existence from anything about him: fingernail, skin, knuckle, whatever. He would know that even he cannot make that change because to change would require a change into its complete opposite and the complete opposite of a speck of God's finger would amount to nothingness, not a finger without God's being. Since the divine finger is already deeply into nothingness (Genesis assumes that God exists adjacent to the void), what would happen to the speck of God's finger? It would float about in empty space, the lost speck of God's finger, but it would remain divine, fully possessed of God's being. Similarly, God would know that the change could not be a change into its almost opposite (as the writer of Genesis believed when he wrote that God had created us in his own image, but not quite divine). For the almost opposite of being is also non-being. All things may be possible to God except the right and the ability to make himself disappear. Where would he disappear to? The void, which has no portal through which he might pass and behind which he could not hide? Being divine, he cannot die; thus everything said to be divine is immortal.

What happens to our mortal souls, the souls God breathed unto us after he had sculpted us out of dust? Not a thing, for we never had a mortal or immortal soul so we had and have nothing to lose. We are either completely divine in nature, and therefore immortal, or we are not divine in nature. The evidence that billions of us have died and that, for the moment, we continue to die denies our divinity and therefore the immortality of our souls as well as bodies.

Besides the spiritual immortality our divine souls would have irretrievably granted us there is the question of why our minds are not equally omniscient with God's mind. Is God's mind possessed of God's being, or is it not? Did the breath he presumably breathed into us possess God's being? Does every speck of God possess his being or is some of God without God's being? In short, is there a portion of God that does not share the exact same being as the rest of God or is it some other creature/entity? No, the Good Book states that God did breathe his breath unto us so we were blessed with God's very own breath, an aspect of God replete with God's being; that, or Genesis was speaking nonsense. So we were endowed with God's mind. But the truth is we can barely see our noses in front of our faces. We are, as a matter of fact, as dense as dogs, according the highest honors to those among us who can predict the course of the stock market or who can predict the possibility that icebergs might be floating in the North Atlantic.

What have we lost? A God in Genesis who vows that we will never be immortal like him and his angels; a God who otherwise makes us in his image, even infusing us with his divine breath, but who nonetheless denies us a place in paradise with him. Worse, a God who capriciously also condemns us to death on earth — body and divine soul — because Eve allowed the serpent (Satan) to talk her into eating the fruit of the Knowledge of Good and Evil. Eve in turn tempted Adam, who shared the joy of first sex with her, ingloriously paying the penalty of work — of having to earn his keep by hard labor ("sweat of his brow") for the entirety of his now abbreviated life. What a monster we have worshipped! As if it were the devil who invented our sexual endowments as well as desires.

A Savagely Cruel God

But wait, Genesis is merely the start of his cruelty. Jehovah condemned us not merely to die but to die savagely and in pain. One wonders how many eons it took He Whose Name Is Unutterable to design Alzheimer's disease; how long before he conjured up Parkinson's; how long to think of multiple sclerosis and cirrhosis of the liver? Was he bored during all those trillions upon trillions of years, with no recreation except to watch his precious angels dance the dance of the seven veils? Imagine having to create the world every fifteen billion years or so, each time hoping to make it better, only to find that the sadistic bent of his finger only made them worse. Did he invent polio, tuberculosis, pneumonia, stroke, gangrene just for us or were they cumulative designs gathered from the infinite number of his creations which he now showered on us? They must be cumulative; no God could be so merciless. O ho, we must remember that he had oodles of time on his hands. What else was there to think about? More earthquakes? No, earthquakes are too impersonal unless you happened to be standing on a fault line. Sadism is best exercised on individuals, for even in great crowds it is the individual who dies, as in yellow fever, malaria, sleeping sickness, crib death, childbirth, smallpox, diphtheria, starvation, cancer of the brain, lungs, eyes, heart, liver, ovaries, prostates, testicles, cervix, colon, throat and skin too long exposed to the sun (who told you to walk around half naked? who told you to smoke?). The list is relentless. Heart attack, heart failure, ventricular fibrillation, faulty heart valves, migraine headache, schizophrenia, depression, blindness, deafness, muteness, infanticide, parricide, matricide, murder by gunshot, murder by strangulation, murder by suffocation, murder by stabbing, murder by immolation, murder by drowning, unaided death by drowning, famine, drought, death by thirst, killer bees, poisonous snakes, crocodiles, alligators, black widow spiders, tigers, lions, panthers, cheetahs, leopards, sharks in their many murderous manifestations (don't swim in the ocean), gallstones, kidney stones, black lung disease, appendicitis, genocide, burning at the stake, shot with arrows, crucifixion right side up and upside down, stretched on the rack, hanging, flaying, the electric chair, gas chamber, potassium chloride, syphilis, gonorrhea, AIDS, and death by dog bite. Or rat bite. Pathologists could surely extend the list.

Did Genesis Create Jehovah in Our Image?

However, it would take an anthropologist to build the list of atrocities *homo sapiens* has visited on his fellow man. Starting somewhere around 400,000 years ago, *sapiens* began the systematic slaughter of Neanderthals, a kindly, beetle-browed species who should really have inherited the earth if Jesus were truly prophetic about the rewards of meekness. By the year 30,000 BC Neanderthal man had all but disappeared; extinct, though some scholars believe that Neanderthal man might have been crossbred into the *sapiens* line, which sounds more like wishful thinking than fact for *sapiens*, it turns out, was more than an aggressor warrior, killing whatever stood in his way; he was a cannibal. What paleontologists have discovered are Neanderthal and *sapiens*' skeletons with holes carved into the bases of their skulls, apparently to make it easier to suck out the brains of their victims.

Whom Can We Blame for Our Mortality?

Is it fair to lay the fact that we have died at the feet of Jehovah/Allah? Whom else should we blame? The writer of Genesis proclaimed that Jehovah had surrounded the Tree of Life lest "they become gods like us." What else could the scoundrel have written? The evidence that people die after four score and ten years was preponderant. Having claimed that God had infused us with his divine breath, he had to explain why our immortal souls had to die with our bodies. The execution of his policy the Almighty left to Gabriel, the angel of death. Presumably, Gabriel takes his orders from God. What choice do we have? Should we blame the devil? Zoroaster refused to believe that Ahura Mazda (Zoroaster's name for God) was evil, so he invented the devil and made him the source of all evil in the world. (Depending on what date you accept for Zoroaster/Zarathustra — 1600 BC or 650 BC — you can date the writing of the Old Testament by it. Conventional wisdom accepts the 650 BC date.) Until Zoroaster, evil, in Semitic theology, was embodied in a little impish creature who snarled his way through the desert camps of the nomads, leaving death in his wake as he scampered by. This little Tasmanian devil was the bogey man who grew up to become Jehovah, a war god, for a while, until he failed, at which point he was discarded and abandoned until he was rescued and re-enthroned by David.

The Redeemer

Jesus was the redeemer sent by God to free us from the sentence of death. The pity is that we continued to die in even greater numbers (millions, perhaps billions, in the cause of Christianity). The redemption of Christianity, therefore, was hedged to allow for immediate immortality of the soul with the promise that our bodies and souls would be reunited on Judgment Day. It was a much sounder theology that Jesus fashioned because it recognized that our immortal souls were divine in nature, but it did nothing to allay the murderous impulses of our human nature. Robert Ardrey called us "killer apes." Given the number who die each day from war, police action, armed robbery, murder for profit and vengeance, infanticide, rape, and pure malice, and auto accidents involving drunken drivers, we might rightly have earned that description for we indeed appear to be driven by the epitaph Genghis Khan is said to have wished carved on his tombstone: "Kill! Kill! Kill!"

Chapter 5. Progressive Evolution

No intellectual achievement surpasses the value of Darwin's theory of evolution for its contribution to human understanding of who we are and how we got here. In effect, Darwin achieved scientifically what the writer of Genesis tried to achieve by guesswork and by fiction.

Natural Selection

Briefly, Darwin discovered that the variation among species (why trees are not like flowers, turtles, rabbits, etc.) was caused by an accumulation of mutations in what he called the "units of heredity" (the word "gene" was not coined until fifty years later), making the individuals blessed with such mutations fitter to endure the struggle for existence. Thus the term "natural selection," which simply means that the mutation(s) happened in the normal course of everyday life and were beneficial.

Good Mutations

They were also mutations over which we have no control. The sun, for example, is responsible for most mutations, good and bad. A good mutation is one which provides an advantage in the struggle to exist, thus "naturally selected." Modern biologists use short-hand language to refer to individuals enjoying good mutations as individuals who are "selected in" while individuals are said to be "selected out" where the mutation results in death or is severely limiting in the struggle for existence (defective genes). Darwin also held that these "transmutations" were inevitably dispersed within a breeding group, thus endowing the group with the beneficial mutations of its individuals. The concept of hereditary units (genes) also identified the process which accounted for improvement of stock at the hands of farmers who raised horses, cattle and sheep.

Lamarck's Theory

Until Darwin published his theory, the prevailing scientific explanation of variation among the species (what caused the variation among species was the centerpiece of scientific intellectualism in the early 1800s) was the theory propounded by the French biologist Jean-Baptiste Lamarck. Lamarck believed that variations among organisms were the result of a process that might be called wish-thinking. A giraffe's long neck, for example, was the result of constantly having to stretch, over thousands of years, to reach the leaves on low-lying tree branches. In effect, Lamarck believed that heredity was shaped by an individual organism's habitual need to adapt to its environment. A bird who continually used his beak to poke for bugs in tree bark eventually developed a chisel-sharp beak and came to be known as a woodpecker. Thanks to Darwin, we know that the opposite is true: mutations shape the beaks of woodpeckers, who then discover that they are able to feed on insects living under tree bark.

The "Invisible Hand"

What troubled Darwin about the concept of natural selection was the implication that something in nature (he was accused of implying an "invisible hand") was hell bent on making some individuals stronger or smarter than others. The invisible hand could be God's, but what a cruel God he would have to have been. While Darwin was not disrespectful of the Almighty, he was certain that the answer to the riddle would be a natural one, not supernatural.

Modern Theory of Natural Selection

In 1925 or thereabouts, a famous research biologist at Columbia University, Thomas Hunt Morgan, amended the theory of natural selection to keep it current with progress in biology. His theory, now called the modern theory of natural selection, states that an individual possessing a gene resistant to a poison or drug which kills off his brethren is said to be naturally selected. We are most familiar with the theory in its application to antibiotics. We say that a colony of bacteria has developed resistance, say, to penicillin, when the fact is that some of the bacteria were blessed with a gene resistant to penicillin. Most of the colony dies but the gifted resisters quickly form a new colony. Guided by this principle, doctors wisely hold back the use of penicillin except in special circumstances, reserving it for extraordinary trauma and for those critical moments in old age when we are likely to develop pneumonia.

Best Genes

New language is constantly finding its way into biology. Several years ago, an anthropologist at Rutgers University coined the phrase "best genes" to define the genes of successful individuals: seven-foot-tall basketball players, champion boxers, Nobel Prize winners, Rhodes scholars, movie stars, scientists like Darwin, Einstein, Watson, Crick. Words like beneficial genes, dominant genes, recessive genes (the latter to be credited to Gregory Mendel) have also become commonplace among academics. Unfortunately, scientists have also come to accept the notion that some mutations are damaging, even fatal, like melanoma, a particularly pernicious form of skin cancer caused by sun damage; this is a difficult concept to accept since the sun is also the source of beneficial mutations.

Progressive Evolution

The discovery of telomeres and telomerase (see below) begs for redefinition of the modern theory of evolution as the mutation of chromosomes in certain individuals of a breeding group which permits them to live longer lives than individuals in other breeding groups. The discovery of telomerase may also permit us to achieve biological immortality, which implies a corresponding increase in the range of the immune system, and a DNA freed from the danger of lethal genes. In effect, an organism headed for perfection; defining a perfect organism as one free of illness, free of disability and free of the threat of what we loosely call natural death. This view of evolution may fairly be called progressive evolution.

Actually, Darwin's definition of evolution implied progression, though without the stated goal of biological immortality. Darwin was satisfied, first, to explain what caused the variation in species and, second, to explain how individuals in a breeding group unconsciously improved their strength and appearance through sexual selection. The preference for tall men and women in a breeding group is a commonplace example of sexual selection as is men's preference for women with large breasts. How breasts get to be large to begin with is the result of natural selection (mutation).

Devolution

The antithesis of progressive evolution might be called devolution, to add yet another term to the art, if not the science. Devolution is the negative mutation of one or more genes in an individual — and ultimately the breeding group — which leads to disease and premature death of the individual and sometimes the extinction of the breeding group.

Static Evolution

Geneticists believe that each of us carries four lethal genes, usually brain genes. This stage of devolution may fairly be termed static evolution, which is the belief that we have attained evolutionary perfection. As evidence we point to our extraordinary intelligence and divine beauty. Those of us so deluded regard death, illness, physical impairment, hurricanes, earthquakes and famine as the work of a cruel nature, made by God or not, but if made by God then made to test us in this vale of tears. We think that the best we can do is to attempt to control nature where we can: fighting disease, lions, tigers, snakes, crocodiles, alligators, tse-tse flies, malaria-breeding mosquitoes, black widow spiders, tarantulas, scorpions and infant-eating rats, provided we put aside the fantasy of life without death because, well, because death is part of life. This group of naysayers includes a materialist intelligentsia (John Dewey might have called them "educated jackasses") which believes that biological immortality is the opiate residue of religion. Evolution has no goal, they like to say; the intellect is supreme and if life needs a goal greater than the continuing enrichment of culture, let that goal be the conquest of space. The smaller number of our brethren who appear not to be caught up in devolution are those who are relatively free of disease and impairment, living well into their nineties and hundreds, and reproducing themselves.

Genetic Determinism

It would appear that our ability to live well into our hundreds has been structured into our telomeres. Recent telomeric research shows that we are born with unequal lengths of telomeres, which probably have accounted for the discrepancy in longevity among great groups of people and even among members of the same family. As we will read below, our telomeric chains lose a section every time our cells double; when the last of the sections are gone, we die. So while we have been correct in blaming many of our diseases and disabilities on defective genes, we have been slightly incorrect when we have attributed senescence, old age, and death to our genes, but grossly incorrect when we assume that this trio of horrors is perfectly natural.

Biological Immortality

What evidence is there that biological immortality is anything more than a pipedream? Darwin had the answer in his hand and let it slip, for it was his discovery that species evolved as the result of an accumulation of beneficial genes which are reproducible thus spreading the benefit throughout the breeding group. Centenarians produce centenarians unless some environmental factor intervenes. The best example is Abkhazia (Russia) which has the highest ratio of hundred -year-old people — and a greater number of them — than any other nation in the world regardless of size. (See *The Methuselah Factor*.) The Abkhazians give credit for their longevity to cold stream bathing. No one doubts that cold water baths help since cold weather as well as cold baths slow down metabolism, thus reducing the number of times the body's cells will reproduce; an important consideration in the now-defunct age of Leonard Hayflick.

The problem is that everyone, or almost everyone, believes that living into one's hundreds is an accident. (Which individuals are hit by a beneficial mutation and which are not is pure chance. However, the ability to use the benefit, that is, a genome ready to make the small benefit work, is the result of an accumulation of genes favoring longevity.) Darwin was insistent that mutations happen slowly (nature doesn't make leaps in the matter, though a Scottish ewe (pre-Dolly) produced a crib full of short-legged lambs which keeps them from hopping over fences into another farmer's flock). Getting to one hundred, therefore, is a slow, arduous process since most centenarians suffer some of the discomforts of aging despite the evidence that people who live into their hundreds are very rarely sick.

Extended Life Spans

If living into one's hundreds depends on the accumulation of beneficial mutations (telomeres), one thing seems certain: the genome must be prepared to use the benefit, which is to say, have in place (inherit) the number of telomeres required to live to one hundred and six, or one hundred and ten; which, in turn, is to say that one or more of your parents and/or grandparents lived to ninety-seven or ninety-nine. The alternative would be freakish; it would imply that evolution does move in leaps and bounds, when it does so only rarely. The current attitude is that all human beings are born with the potential to live a hundred years or more but somehow lose the advantage; a regression in evolution rather than a progression; in effect, unraveling; going backward; devolution in extremis. A fatal mutation or an accident in everyday life might account for the premature death of those individuals slated to become centenarians (by virtue of family history), but that does not allow us to presume that all individuals are destined to become centenarians or that all men are predisposed to live to 75 or 78 and women to 78 or 81 (actuarial estimates of life expectancy). Actuarial tables are statistical averages. They might justify loose predictions of life expectancy for certain breeding groups but always with the caveat that averages are made of lower numbers as well as higher. (All right if you're in the life insurance or mortuary business but not if you're in microbiological research.) The rule would seem to be that an individual will live as long as his parents or grandparents; or perhaps as long as a long-lived aunt or uncle; we might add a few years or decrease them to account for variations in lifestyles. But the idea that each of us is born with the opportunity to live to one hundred and twenty or twenty-five years provided we follow the rules for good health is not supported by the evidence. A French woman, Jeanne Louise Clement, reached the venerable age of one hundred twenty-two and she smoked two cigarettes a day until her death in 1997. This is not to suggest that you should start smoking again but rather that longevity is deterministic.

Nature Does Not Leap

Unless we are willing to accept that nature does make mutational leaps we have to deny that our forebears on the evolutionary tree were centenarians and accept the doctrine that longevity is the result of a slow accumulation of beneficial mutations. If all of your known ancestors and/or relatives died in their fifties, sixties and seventies it seems fairly certain that you will not reach one hundred or one hundred and ten.

Same rule: nature does not make leaps. Find your consolation in the highly probable circumstance that your ancestors were lucky if they lived to fifty. If your ancestry is European, you should also celebrate the fact that your ancestors escaped the bubonic plague which obliterated three-fourths of Europe's population in the fourteenth and fifteenth centuries. They had a gene which was resistant to the bubonic plague and you probably still have it. Fortunately, all this is about to change.

Lucy

When you track backward in history with biologists, archaeologists, and paleontologists to guide you, you will reach a point where there are no fossilized skeletons of human beings with which to make comparisons regarding age. Richard Leakey discovered a tiny two-legged simian creature he lovingly christened "Lucy," whose date he estimated to be 3 million BC. Let's be generous and say that Lucy lived to be twenty-five. In three million years Lucy accumulated beneficial mutations which allow her descendants — us — to live to 122 years (Jeanne Louise Clement). They — the accumulated beneficial mutations since Lucy — have also allowed 450,000 people around the world to live into their hundreds

Could Lucy have been older than twenty-five? Yes, but what difference would it make? One could argue that Lucy might not have been the ancestor of humankind. Indeed, but what of all the fossilized human skeletons uncovered by archaeologists whose bones reveal that none was older than sixty at the time of death? You could also believe that the original woman was a gorgeous twenty-three-year-old married to a handsome thirty-year-old, both of whom lived into their sixties. In short, you could continue to believe in Adam and Eve but you would still have to account for how 450,000 of our world's citizens are living into their hundreds and beyond. You could point to Methuselah (a great example of devolution *in extrem0is*) or you could accept Darwin's theory of the descent of accumulated mutations which we now know favor longer and longer lives (biologists estimate that there are fifty thousand to one hundred thousand genes in every cell).

This Cycle of Evolution Still Incomplete

Without Darwin's belief in the accumulation and descent of naturally selected mutations, nature would be in chaos; devolution would be the driving force, not evolution. It would be tantamount to saying that organisms pop up out of the muck only to disintegrate. The planet would be a constant display of bio-cosmic fireworks, which is pure nonsense. Organisms have to evolve in order for them to devolve; not because devolution is the primary driving force, the "invisible hand" moving species forward but because the saga of evolution is not complete; far from it. We progress but, for the moment, we also die. We also reproduce so that the beneficial genes we have accumulated may accumulate further; slowly, agonizingly slowly, in a multimillion-year drama which has repeated itself, and will repeat itself, an infinite number of times.

Progressive evolution is like the rising tide reaching higher and higher on a beach, then leaving unequal undulations of foam as it recedes. Intelligence, beauty, and size have been for us human beings the most exhilarating tracings on the shore but we have hardly begun the arduously slow process of evolution. We are porters in the process, carrying for our children and future generations the blessed accumulation of beneficial genes upon which to build the edifice of biological immortality.

Evidence of Evolutionary Immortality

Putting aside the discovery of telomeres and telomerase, where is the hard-edged scientific evidence that evolution is headed toward biological immortality anyway? The best evidence that nature is headed toward the goal of continuous life is the human immune system, which is reinforced in the non-mammalian world by creatures that grow new limbs when they lose them. Operating wholly independently of our "divine" minds (which look stupid by comparison), the human immune system protects us against all attempts by foreign matter to damage any part of our bodies and it has evolved so intelligently that it makes copies of such enemies so as to minimize any future assault by the same invaders (which is how and why vaccines work). The immune system has the goal of maintaining the integrity of the organism against attack by disease carried by viruses and bacteria and by external and internal injury, which is another way of saying the prevention of death. Our skins are another form of defense against the microbe world. Almost as impressive as our immune system is how evolution has developed scar tissue to heal wounds. Not bullet holes or deep gashes from which our lives gush away in uncontrolled bleeding, but the hundreds of everyday wounds which quickly seal the inside of our bodies and stanch the flow of blood. Let us not forget the healing function by which our broken bones stitch themselves together. Yes, they will set crookedly, perhaps, if left to their own devices, but who can doubt that, like a penguin's flippers, evolution will one day provide the mechanism which will set them correctly? (Of penguin's flippers Darwin once asked, who can doubt that one day they will grow into wings?)

The fact that orthopedic surgeons can set bones correctly and that general surgeons can stitch flesh which has been deeply traumatized is a testament to the precocious evolution of the human brain. Medical science has anticipated much of what evolution will probably develop anyway. We call it discovery of nature's secrets and term the process scientific research. The evolution of our super brains has allowed us to build huge arsenals of medicines and instruments whereby our physicians and surgeons cheat death every day. Indeed, the *raison d'être* of medical science might easily be said to be the prevention of death and of life-threatening injury. Cultures, beginning with religion, have taught us to accept the inevitability of death and to focus either on preparing for the hereafter or for living life to the fullest while we are here, or both. Living to the fullest while we are here leads to the worship of money and money-making, which ultimately leads to the competition for who is the richest man or woman alive, which is a form of fame, which in turn promises inclusion in the history books and historical immortality. There is no escaping the compulsion to be continuous for we are made of continuous stuff which colors everything we do. In the view of progressive evolution we might even dare say that the stuff we're made of has been here before: ten thousand, twenty thousand, an infinite number of times. We are about to fulfill our destiny once again and to repeat the process an infinite number of times in the future.

Does anyone who believes Darwin's theory believe that life on the planet began with hundred-year-olds? Or fifty-year-olds? Or ten-year-olds? Life began as Darwin surmised with unicellular animals who, as the earth cooled and oceans formed, mutated into the more complex organisms of marine life. Our origins are in the sea. Sharks were fully formed as they appear today, chinless and nasty, 600 million years ago, eons before the first dinosaur was a mutated mote in some reptile's slitty eye. We have little choice but to believe that something changed them from single cell organisms to multicellular organisms. The alternative would be oceans chock full of never-changing single-cell animals. To the contrary, what we have are highly complex organisms displaying a near infinite variety of forms, all struggling to survive in the forms in which they find themselves and all but a few failing.

The accumulation of beneficial genes in some marine animals who, testing the shores of the receding oceans to find that they had developed lungs, eventually resulted in the massive exodus from the sea of amphibious creatures who thereby laid the foundations of all land animals. This march from the sea onto land is a breath-taking panorama which never fails to inspire us with pride in the inquisitive and intrepid nature of our evolutionary forebears. Ask not therefore why mariners dare to chart unexplored waters; neither ask why we need to climb the highest mountains or to walk on the moon.

Chapter 6. Relative Immortality

Based on paleontological and anthropological evidence, it would appear that the atom's development as a reproducing and reasoning entity has been a long, arduous process, with chance thrown in to explain why it resulted in us. Unless astrophysicists are holding back evidence (not likely) it is, of course, impossible to say, and foolhardy to guess, that perhaps sometime in the last fifteen billion years the atom did fully realize itself as a reproducing and thinking entity on some unknown and undiscovered planet somewhere in the splendor of galloping galaxies ahead or behind us in time. All we know is that, here on our planet, the process has been long and arduously slow.

It has also been stubborn. The compensation for this excruciatingly slow process of evolution is a wondrous tenacity that belies the appearance and disappearance of the billions of other species whose skeletons lie preserved beneath the crushing strata of cosmic dust. The shark exists pretty much as it did 600 million years ago. The Virginia opossum is structured with the same skeleton it boasted 400 million years ago. This is not to say that the shark and the opossum are immortal but rather that their species are immortal (or, in biologist's terminology, species-specific). The redwood trees of California are a thousand years old. The giant tortoises of the Galapagos purportedly live to 180. And we, when we're lucky, to 122 years.

All Living Things Are Headed for Immortality

The tenacity of longevity might appear to be governed purely by chance (whose mutations are beneficial, whose not) but there is the possibility, rarely considered by biologists and other evolutionists, that all living things are headed for biological immortality. (We will have to hurry; astrophysicists estimate that the life of our sun might be limited to only another five billion years; but more later about how foolish we sound when we talk about a billion years as though it were a century or a millennium.) It is a pity that the billions of extinct species that are buried in the bowels of the earth will not be here to celebrate that wondrous day when statisticians will predict as a matter of their daily reportage that the life expectancy for women is 822 and for men 804, the way they now forecast life expectancy as 78 for women and 74 for men. It is pure speculation, of course, that evolution is aiming toward our immortality, but given that three billion years ago, or so, most of our planet was muck and water, we must admit that we have come a long, long way; from tiny unicellular organisms to monsters like the dinosaurs and whales and mammoths.

The Sound of Our Atoms

Immortality sounds too much like the wild aspirations of religious fanatics to merit weight and dignity as a possibility in the halls of science (but perceptions are rapidly changing). It seldom occurs to some of our scientific brethren that the craving for immortality is the sound of our atoms screaming for attention: listen, stupid, you're made of immortal stuff, you're made of atoms that never die, not even in the cauldron of the sun; never mind that people and other animals die every day; focus on the fact that the atoms we're made of don't die. Ever.

We'd much rather focus on the disappearance of things, building large pools of water underground so that we may observe and measure the loss of a proton. Just think — if we can prove the absolute loss of a single proton we can then predict the death of the earth and, perhaps, the death of the universe to the day; well, maybe not the day, but perhaps the century. What a pity.

The universe isn't going anywhere. Nor did it come from anywhere. Something cannot become nothing. Nothing cannot become something. The universe did not begin, let alone begin as an incredibly dense pinpoint dot emerging from no one dares say what.

The universe always has been and always will be, which is to say that the stuff we're made of is immortal in the most literal sense of the word. The atoms in our brain know it which is why they scream bloody murder when we're hurt — how dare you damage this immortal body? — and why they sob and mourn when we bury our dead.

It is comforting and often rewarding to talk to your neurons and to let them know that you're aware that: (1) they are immortal (2) they dread dispersion and (3) you're doing the best you can to keep that from happening. Which is, in effect, what belief in the immortality of the soul has done, lo, these thousands years, and why we have fought so fiercely to defend that belief.

The Hayflick Limit

In the 1940s, Alexis Carrel, famed French biologist and 1912 Nobel Prize winner in Medicine, announced that experiments he had conducted at the Rockefeller Institute in New York had demonstrated that normal human cells were immortal. By this he meant that the cells continued to double when they were supposed to be dead. The reaction was electrifying. After all, continued doubling of cells that were supposed to be dead meant that what we call "natural death" might not be due to inherent breakdown of the body but might be due to some unknown environmental factor.

Carrel's findings held until 1961 when Leonard Hayflick, noted microbiologist, rained on Carrel's parade by announcing that his own attempts to replicate Carrel's findings had revealed a limit of approximately fifty doublings in normal human cells. Hayflick guessed that workers feeding Carrel's petri dishes had inadvertently introduced protein into the dishes.

Hayflick was immediately challenged by respected microbiologists — notably Bernard F. Strehler — who argued that human cells double thousands of times in a lifetime. But the damage was done. The Hayflick Limit became the battle cry that dominated cellular research for the next thirty years. However, it was a battle cry that was muted by clinical evidence that the Hayflick Limit did not apply to cancer cells.

Henrietta's Cells

Eight years earlier, in 1953, a woman named Henrietta Lacks had checked into Johns Hopkins Hospital in Baltimore with what turned out to be an incurable case of cervical cancer. Henrietta eventually died but not before her cancerous cells, happily doubling in petri dishes at Johns Hopkins research laboratory, demonstrated an indefatigable need to continue doubling, thus earning the sobriquet "immortal." Henrietta's cells became international celebrities. Now called *Helacyton gartieri*, the direct lines of her cancerous cells provide the basis for study and testing in laboratories all over the world. And they're still doubling. Biologists estimate that Henrietta's cells will continue to double for "thousands of years."

The Discovery of Telomeres

In 1998, a stubborn but stalwart team of American cell biologists demonstrated that the induction of a chain of enzymes called telomerase into normal body cells will prevent cellular death, in effect making us immortal. The study (Bodnar, 16 July, 1998, *Science*) was based on the earlier discovery by Howard Smith (1985) that attached to both ends of each of our 46 chromosomes is a single strand of DNA divided into sections called telomeres. Every time our cells double the strand loses a section of telomeres. As the strands run out we grow old, when the strands are used up we die. What the Bodnar study showed was that when our chromosomes were capped with telomerase the telomeres remained stable. In effect, our telomeric strands are a tick-tock factor measuring out the length of our lives. Since we are born with telomeric strands of different lengths, we die at different ages.

The Bodnar team had accepted a formidable challenge. The fear among cell biologists was that telomerase was a marker for cancer, or worse, the cause of cancer. The only beneficent implication was that, by keeping the telomeres intact, telomerase permitted the cancer cells to continue their murderous doublings. What would happen when normal body cells were inoculated with telomerase? What happened was the conquest of death.

Howard Smith and Andreas Bodnar's team were not the discoverers of telomeres and telomerase which were known to cell biologists from the 1950s. What Smith and Bodnar discovered was the connection between telomeres and what we call natural death, and the connection between telomerase and cellular immortality.

The Evolution of Telomeres

Have our telomeres evolved or were they always there in the same number? For those of us who believe in evolution the answer is, yes, they have evolved, meaning that at one point in our evolution we had a smaller number of telomeres pinned to our chromosome ends. The alternatives are to believe that: (a) telomeres are not the tick-tock factors of life and death, or (b) deny evolution, which implies that we were created pretty much as the Good Book says.

Since by now the evidence supporting evolution is overwhelming, it appears reasonable to assume that telomeres have evolved, which automatically means they are subject to mutation, which is how evolution happens. For the fortunate among our progeny, it means that our telomeres will increase in number, thereby automatically increasing life spans both for us and for our animal friends. Given the lazy clock nature lives by, it is therefore reasonable to infer that evolution will eventually produce species which are biologically immortal.

We need not slip into Lamarckian heresy to acknowledge that the atom is hell bent on making human beings immortal, meaning we have wanted to become immortal and we have finally succeeded. We need only acknowledge that this is what the atom does: in infinite ways it simply expresses its compulsive continuity. When we scoff at the attribution of thinking power to the atom it is because we still assume that "thinking" is an attribute of mind. But without the possibility of mind's co-existence with matter (see Falling Trees, above) we have no choice but to credit thinking to the brain.

Continuous Things Are Compulsively Continuous

Farfetched though it may seem, the concept of compulsive continuity is truly simple: continuous things cannot be anything other than continuous, nor can they do anything that would be more or less than continuous. Continuity is all. What appears to be happening in evolution is a long and arduous process of forming organisms which are, in their totality, as continuous as their simplest components. Another way to put it, of course, is to state that if microbiologists never lifted another finger, all living organisms would eventually achieve biological immortality. It might take another million or ten million years (viruses, bacteria and fungi have already achieved immortality) but it will happen just as surely as the belief that someone in Abkhazia or France will celebrate his or her 150th birthday sometime in this century or the next without scientific aid or interference. But aid and interference are precisely what is expected of the atom in the form of scientific brainpower. So, cast all sheets to the wind, let microbiologists interfere; interfere all they wish. Experiments since Bodnar in the 1990s have focused on a variety of illnesses that may affect or be affected by telomeres and telomerase, chief among them cancer and diabetes. Studies have also included the possible influence telomeres and telomerase might have on stem cells themselves. Tucked away in secret niches in the body is a veritable treasure trove of stem cells which are eagerly responding to the response of telomerase. We may yet live to greet the day when our limbs and organs regenerate themselves as silently and as matter of factly as a lizard replaces a lost leg.

Let Us Not Forget the Naysayers

The naysayers among us, and they are legion among the literati, are quick to remind us that immortality is a short term illusion in a universe which is flying outward in space at speeds approaching the speed of light. At some point, five or ten billion years from now, everything will have been reduced to the thinnest gas (hydrogen) and will begin to reverse itself; or the universe will pull up and pull back, thanks to gravity. What will happen then, heaven only knows, but we will be long gone. The stars will have dissipated their energizing atoms and our descendants will weep as they witness the self-destruction of their immortal bodies.

Not to worry. We will have turned to ice with the death of our sun five billion years from now. However, astrophysicists are not in agreement about the inevitable death of the sun; there are some who would argue that the sun seems to behave like a breeder reactor, constantly regenerating itself. So there is hope, contentious though it be, that our sun may not be doomed in five billion years but will suffer the fate of similar stars. One way or another, crow the naysayers, we are doomed. So sit down and write a Book of the Dead for our own era, give up childish dreams of immortality, and write new slogans to reinforce the ones we love so much like the inevitability of death and taxes. Or, death is part of life, son, so don't be a cry baby.

Unless some worthy astrophysicist offers another explanation for our Valkyrien ride outward into space, we are indeed doomed to inevitable destruction five or ten billon years from now. Give up all hope of being absolutely immortal (except in heaven) and accept the fact that we cannot hope for anything more than relative immortality. Prudence might dictate that we aspire to no more than three or four billion years, since the sun should have begun to cool off eons before its ultimate demise. A truly conservative outlook might urge us to count on no more than, say, one or two billion years of deathless existence.

Relative Immortality: A Reasonable Expectation

The only "forever" immortality we've ever dreamed of has been the religious hope that, come Judgment Day, our souls would be reunited with our bodies to dwell forever in some undisclosed Eden which for many of us has meant a continuation of the life we were living the day death did so brutally sever us from it. It would be fair to say that most of us would consider a thousand years a reasonable definition of immortality. No? Well, perhaps five thousand. Still not enough? Ten thousand, then. Ten thousand years is a long time. Three or four thousand years longer than recorded history. Twice as long as the Colossi in the desert have existed. Three times longer than the oldest pyramid. Five times longer than the entire span of Christianity. We might, after ten thousand years, begin to face a problem almost as forbidding as death: boredom. Never? Are we certain we can stay busy and enough involved not to be bored with living? Yes, yes, yes. Why, it would take thousands of years just to explore our solar system, let alone look for and settle another planet. And if we did not find one that we could easily make habitable and had to escape our solar system altogether, ten thousand years would be a paltry measure — even were we to take into account Einstein's prediction that we do not age when we travel at celestial speeds.

25,000 Years

Still, if interplanetary and intergalactic travel are to substitute for our intercontinental peregrinations, a lifetime of twenty-five thousand years might be warranted. Twenty-five thousand years might also be necessary if one of our goals, say, were to make love to every woman on earth or, for women, to make love to every man on earth. If the adult world population were ten billion and we limited ourselves to one-week affairs, we would need five billion weeks for all living adults to mate with every other man or woman, Five billion week-long affairs would require a lifetime of almost one hundred million years, not twenty-five thousand. If we were less ambitious and not so easily bored, we could stay married to one woman or man for a hundred years and remarry, let us say, ten times, each time for one hundred years. Assuming that we would marry the first time at age 25, we would need 1,025 years to meet our matrimonial goals. What would we do for the other 23,075 years, stay unmarried? Not likely, fornicators that we are. We would have to set different matrimonial goals; say, get married one hundred times for one hundred years each time, which would use up ten thousand years. Or if we really liked changing wives and husbands all that much, we could remarry five hundred times for fifty years each time, which would use up almost all of our projected lifespan of twenty-five thousand years. At that juncture we would not present ourselves to the family doctor for our centennial injection of telomerase. We would declare that we had had enough of living forever.

A life expectancy of twenty-five thousand years hardly qualifies as immortality and seems stingy as relative immortality when the sun has billions of years to go before chilling out. But what would we do if we had to keep reconstituting our telomerase every thousand years or so? Why, we would do it, of course. We might even opt to extend our options beyond twenty-five thousand years. Why give up the joy of basking in the light and the heat of the sun; counting the stars at night. Or returning to lovers whose magic continued to haunt our memories as well as our loins. We might even take up a succession of careers: things we've always wished we could do; people we wished we might have become: doctors, lawyers, judges, teachers, writers, painters, magicians, actors, directors, baseball, basketball, soccer, football, tennis players; wrestlers, boxers. Well, maybe not footballers and boxers, unless we will have reached the point of stem cell repair of spinal cords, limbs, and brains.

With Malice Aforethought

The point is not whether or not twenty-five thousand or one hundred million years qualifies as relative immortality but rather whether or not long extensions of life are endurable and, almost as significant, whether or not the attempt to close off any hope of immortality, relative or absolute, is inspired by malice. How else are we to describe the long-held frantic need to deny the possibility of immortality and to affirm the Good Book in which God posted archangels around the Tree of Life "lest they [Adam and Eve] become gods like us"? God might have changed his mind.

Christians believe he did change his mind. But it is not Christians or Hindus who predict the end of the universe five or ten billion years from now to argue the futility of immortality. It is a band of old warriors, about to wave goodbye, who attempt to justify the destructive nature of the old materialism. After all, the purpose of the scientific revolution, scientific method, and scientific materialism was to rid us of the poetic errors of religionists and the cockeyed speculations of Aristotle about nature and natural law. Are we now to affirm, by implication, that the ancient Egyptians were right; that their dreams of immortality were the images relentlessly forged by the atoms in their brains which came into consciousness through the symbols of their time: a ride in the eternal chariot of the sun? Not a bad guess, as it turns out. Fifteen billion years, with five billion to go, is indeed an eternity.

The argument that immortality is an illusion because the universe is going to hell in a hand basket amounts to a red herring waved gloatingly by intellectual delinquents who would kick God in the shins if they had to, if only they could rid the world of the word "immortality." Immortality, they believe, the hope of and need for it, is what keeps the dreaded enemy of religion alive and prosperous.

Where the argument against immortality is mounted by non-intellectuals, what we hear instead is: where will we put all the people? The assumption is that though we might have achieved immortality we would continue to reproduce at the same rates as before: in some countries, as in Latin America and India, doubling populations every twenty-five or fifty years; in others, like Italy, once the kingfish of reproduction, now doubling only once in every one hundred and eighty years; in still others, like China, actually reversing population growth. (One child per couple will produce the opposite effect of doubling: halving).

An equally valid assumption is that the attainment of relative immortality will dampen the compulsion to reproduce. Reproduction is simply one way the atom has found to express what it is: a relentlessly continuous entity which will express its compulsive continuity in as many ways as may be necessary simply to be continuous. Since the atom appears to be endlessly supple and creative, we should not be surprised if trees began to bear a new race of children, like oranges; not likely, of course, but we must be ever vigilant. Such trees would have to be uprooted as would plants bearing children like cantaloupes and watermelons. We could remind the atom that, having split it once and fused it once, each time forcing it to disgorge its energy, we might do worse the next time. The atom, commenting as a laughing hyena, baboon, or ordinary human being would merely guffaw: let the human species join the dinosaurs if it will; the atom will begin again or take up with some species which is at this very moment standing in the wings waiting to replace us. The wiser atoms among us will not permit the extinction of our species. Like the Chinese, they will decree a limitation on family size when and if necessary. A daily, weekly, monthly or annual lottery may be a voluntary solution. The winners would earn the right to reproduce themselves — once. The danger of over-population is not an obstacle to the enjoyment of relative immortality. Where we would put all the people is simply another red herring.

Would cloning be a reasonable alternative? To what? Reproduction? Not very likely, since cloning is a form of reproduction. Indeed, from the viewpoint of population expansion reproductive cloning (as opposed to therapeutic cloning) would be a disastrous alternative. By limiting sexual reproduction to one offspring per couple we would reduce the population by half in each generation. Cloning, which would allow one clone per individual, would maintain a static population once everyone had been cloned. The primary purpose of reproductive cloning would be to obtain nearly exact copies of some extraordinary individual. Allowing everyone to clone would be democratic but pointless except as an exercise in egocentricity. Besides, in a law limiting reproduction to one child per couple, whose body would be cloned? Mom's or Dad's? No, cloning is clearly a strategy for dying populations where we cannot bear the idea that we would lose forever the alluring charm of a great beauty or the awesome inventiveness of a scientific genius. Once the sanction against reproductive cloning were lifted, cloning might then be regarded as an alternative to sexual reproduction. After all, halving populations would, in three or four generations, become a self-defeating strategy. Someone has to grow food, practice medicine, pull teeth, police the streets, nurse, cook and so on.

From none of the above should we infer that a carbon atom, say, reproduces itself like a virus or bacterium; nor should we infer that reproduction means the creation of more carbon atoms out of nothing. To the contrary, the atom appears to remain as determinedly impregnable as we have found it to be and infinitely destructive when, acting as thinking atoms, we invade its tight construction.

A Sin Against DNA?

Geneticists teach us that when DNA uncoils to permit one helix to create a template in the nucleus, the product of the template (RNA) then hurries out of the nucleus, carrying its message to the cytoplasm where tons of amino acids, the products of nutrition, wait to be selected. RNA picks and chooses the precise amino acids required to make an exact copy of that particular cell. RNA does not create amino acids;

for that matter, it doesn't create anything except in the loosest sense of the word. What RNA does is to assemble the amino acids in the order dictated by DNA.

Will we be committing a sin against nature by curtailing reproduction in an immortal society? Not very likely. The sin, if we may call it that, would be to deprive another human being of what would certainly be a natural right. In natural law, a sin is anything which promotes the death of an organism. Considering how long it took us to attain the forms we now enjoy, the sin would indeed be unforgivable.

We rarely stop to remember that we have been, in fact, immortal until now. Our parents, their parents, grandparents, each generation of our progenitors had to have been alive when it reproduced itself. Our gametes (sex cells) have never died nor have the beneficial mutations we have accumulated since emerging from the slime died. We who live and breathe are immortals. We sneer at the designation because we cannot reach for a stream of memories that would evoke events in the long ago past. That's what we mean by being immortal. Indeed. Most of us can't remember what we did at age five, let alone ages two, one and prenatal. But we were alive. Nor are we willing to accept that our ancient memories are stored as muted reflex behavior: how to smile, cry, see, hear, speak, flail our arms, crawl, stand, walk, sleep, awake, suckle. It took millions of generations to acquire those talents. Imagine how exciting it must have been when we were first able to stand or shout. None of that counts now: it's much too dumb, too hushed, too instinctive. We are super-conscious organisms now. We require a long stream of unbroken consciousness to qualify as the condition of immortality. Unbroken, back at least, until we were seven or eight.

This is not to imply that, with immortality, we will lose our long term memories. It merely assumes that the memory process we experience now in our abbreviated lives will or, rather, may be mirrored in our prolonged lives by the loss of short term memory and retention of long term memory, up to a point. With prolongation of life we may also gain greater memory storage capacity or, with our growing dependence on computer memory, we may lose memory storage capacity (use or lose). How big a brain do we really want to carry on our skinny necks and narrow shoulders? Gratifying as is the retention of long term memory and the gradual loss of short term memory, evolution has adapted our memory storage capacity to our needs. It has recognized that we needed to retain as much information as we dared cram into our early years, then selectively it sloughed off the slag heaps of information with which we are bombarded in our later years. Our brains are smarter than we think they are. Whether or not any of this speculation can be proved is a matter for brain specialists who may or may not cater to our petulant demand for perfect memories. What is relevant with the achievement of enormously prolonged lifetimes is the enjoyment of them and the security of knowing that we can continue to extend them as long as we wish. It is a far cry from trying to decide whether or not to request assisted suicide.

The Sinful Option of Suicide

Nothing in the promise and process of long extended life precludes the option of suicide. One assumes that the pains of aging for which suicide seems to provide the only escape will disappear. It is a fair assumption: no aging, no pain of Alzheimer's. But what of the proliferating rates of teen-age suicide in countries like the United States and Sweden? The urge to commit suicide, flying as it does in the face of compulsive continuity is, to borrow a phrase from Winston Churchill, an enigma wrapped in a riddle. If the prospect of a thousand or more lifetimes does not give a young man or young woman reason to live, the extended span should at the very least give psychiatrists time to examine the reasons fomenting such pointless acts of self-destruction.

One prays that with biological immortality the old habits of reaching for substitute forms of immortality will also disappear. We can only caution the young man or woman who has become desperate because he or she believes he or she is not beautiful enough or talented enough to become a movie star or rock star and therefore seeks high-profile singularity through suicide: a thousand lifetimes may reveal talents you never suspected that you had; there is plenty of time to end one's life. It has taken a billion years to get this far. Why discard it before you're twenty for the dubious and fleeting distinction of having dared to do what others will not: leap from a twenty-story ledge or hang yourself from a beam? Be assured that three years later no one will remember the deed, let alone your name. Where they do, rest assured that all they will remember is your inability to cope. No one other than your immediate family really cares whether or not you're dead or alive. If you were alive instead of dead, you might argue that it takes great courage to take one's life. Suicide requires courage, all right, but it is the courage of those who lacked the strength to deal with life; the courage of the foolhardy, the damned, the weak. Were it not so macabre we might establish suicide parlors where the desperate might choose from a variety of life-terminating options and where their imminent departure would be hailed by flashing reminders that no one cares and by loudspeakers barking the mocking laughter of demons.

Teen-Age Violence

The recent spate of high school violence in the United States has alarmingly revealed that the pressures of being a teenager in America are unbearable for some of our children. We blame gloating athletes unaware that braggadocio can be sadistic if it uses a less talented schoolmate as a foil; or we blame careless fathers for keeping loaded guns around the house. We need also to see that in keenly competitive societies like America's, the Constitution does not shield us from the strutting rights of those whom chance or eugenics has declared the winners in the competition of life. Only in America would a mother hire an assassin to kill a young girl who might have bested her daughter for a place on the cheerleader squad. The culture of immortality will totally transform existing cultures which are now driven by a race against death to determine who has accumulated the most toys, the most money, the most clippings, and the most covers on *Time, Newsweek, Vanity Fair, Entertainment Weekly*.

Mister and Ms. Wonderful

We would do well to soften our cults of personality, and we might begin by eliminating the practice of choosing valedictorians and salutatorians by grade averages. High grade averages speak for themselves. The statistics are always available to those who need to know them. It would also help to eliminate the National Honor Society, Phi Beta

Kappa, Mensa, awards of athletic letters, fraternities and sororities, everything and anything that would reinforce recognition for the gifted and anonymity for the losers. We might also plead with the media not to wallow in the mass murders of the young or to glorify the success of the young in business by publishing their photographs and notices of their promotions, thereby grinding the noses of the less successful in the manure of publicity handouts. Notice, if you will, that with the exception of heroic firemen and policemen, the honorees on the local achievers' page are usually employees of businesses, businesses that are advertisers or potential advertisers. Well, say our presidents, the business of America is business. It is not. The business of America is freedom. Freedom to worship as we choose. Freedom to assemble with whom and where we choose. Freedom to speak our opinions, publicly and privately. Freedom to print what we choose to print. Freedom from want. And now freedom from death.

Freedom from Death

In a way, our governments and hospitals do now provide a certain freedom from death. In most of the civilized world, no one is left to die in the streets. Emergency rooms and ambulances are available to all, the indigent and uninsured as well as the affluent and the well insured. The freedom from death which we need now is public policy dedicated to the proposition that what we call "natural death" is an illusion (or, as some prudent politician might describe it, "may not be necessary or natural").

As it stands now, the ability of science to prevent the attrition of our telomeres is a private matter which sooner or later must become public policy as government elects to decide who will and who will not be entitled to telomerase therapy. Clearly, leaving telomerase therapy in the private sector would be limiting its availability to people who can afford it. If Viagra cost $10 per caplet at its introduction, what would a telomerase injection cost, one treatment of which could extend life for a hundred years or so? Ten thousand dollars if the procedure requires hospitalization, a hundred dollars a shot if it is available at your doctor's office. Beyond the cost, imagine the crush of people demanding to be first. Government intervention would certainly be justified then, wouldn't it? It probably would; but it is hardly an absolute certainty. Why not? Because large numbers of people might resist the government's support of a procedure or treatment that would prevent "natural death." My God, you might think, is there anyone so masochistic as to refuse deliverance from death? Yes, there is. Conditioned to believe that heaven is the only absolutely certain deliverance from death, millions of people might resist the whole idea of biological immortality, urging the government not to support research or treatment with public money. Absurd, you might think. But consider the possibility that religious leaders might pressure politicians, using the rationale that the procedure should not be offered to some until it can be offered to all. It would be a stratagem, a ruse, of course, but we have seen the same political logic applied to human cloning and to stem cell therapy.

One could reasonably argue that reproductive cloning can hardly be compared to a medical procedure that prevents death. But the ease with which "vested interests" persuaded Congress and Parliament to proscribe cloning should be a warning that politicians might be similarly influenced to ban biological immortality. Which vested interest might they be? The political left, in the matter of cloning. The religious right would clearly be the vested interest opposing biological immortality. Why? Because biological immortality would deny spiritual immortality which is the foundation of Christianity, Islam and Hinduism.

Resistance of the Religious Right

How long would the religious right persist in its formal resistance to biological immortality? As long as enough of its adherents continue to support the clergy, the churches, mosques and ashrams. If history offers any guidance, as long as there are people who want to, or need to, believe in life after death. After all, biological immortality is no guarantee against accidental death. An accommodation of some sort may be the way the clergy finally resolves the dilemma. Until then, expect plenty of resistance with most of it deployed in the halls of Congress and the parliaments of the world.

The presumptive resistance of the religious right to the advent of biological immortality might require a like response from the political center, a counter-effort by immortalists to amend the Constitution to guarantee freedom from death and to require the federal government to do all in its power to make biological immortality available to all who would desire it, even as the government assumed administration of the polio and smallpox vaccines, and research into AIDS and other epidemiological threats to life. There is little question that the federal government will find itself in the jaws of a vise. Eventually, it will respond, as politicians are supposed to respond in a democracy, to the will of the people. Freedom from death, then, should be a slogan, a battle cry intended to put iron in the federal government's spine while it holds the religious right at bay. In short, do not take it for granted that everyone will immediately welcome long-lasting life with open arms. Be prepared for resistance, and when resistance fails, be ready with forgiveness.

Chapter 7. Thou Shalt Not Kill

We live on the cusp of the most frustrating anomaly in history. Ludicrous is perhaps a better word. The dreaded fear that we might become immortal while we're still killing each other in wars and revolutions, by terrorism, murder and suicide, is now a reality. What madness! How human! The paramount ethical rule in an immortalist culture must be: thou shalt not kill. Thyself or anyone else.

One might argue that the prohibition against killing is already universal, but that is not strictly speaking true. The prohibition against one-on-one personal killing may be universal but killing in war, revolution, coup d'états, etc. is not. Indeed, such killing is generally promoted as a "good thing," the continuity of nation, tribe, social group being favored over the continuity of the individual person (soldier, air raid victim, and victims of other bombardment). So we have our work cut out for us.

The Ultimate Deterrent to Killing

The prospect of immortality may in and of itself provide a "think-twice" motive before committing one-on-one murder (what fool would give up the prospect of living for hundreds, perhaps thousands of years, for the deranged need to kill someone?). We cannot be certain, however, that it will automatically provide a deterrent to war, terrorism and revolution. Why not? Because the fomenters of wars are usually leaders of the state, whose edicts for and against war are reinforced by the police power of the state. The police power of the state leaves the individual with little or no choice: escape to a country which has foresworn war (Sweden, Switzerland), go to prison, or permanently give up citizenship in the country one has refused to serve by risking his or her life in war. The goal, therefore, in the effort to prohibit killing by war is to prevent the election or the coming to power of swashbucklers who would achieve old-style historical immortality by avenging ancient tribal or national grievances, trumped up or real.

Socrates said that politicians lie (for the good of the state, he assumed) and therefore cannot be taken at their word when they say that they are, and will be, pacifists come what may. What should we do when one of them changes his tune? Assassinate him? Not when we have vowed never to kill. What then? Rely on the police power to arrest and detain him, and then rely on the judicial system to detain him permanently. Would a politician risk confinement in prison for eternity simply to make war? And why wouldn't confinement for eternity be considered cruel and unusual punishment? It probably would, so we would be obliged to modify laws barring cruel and unusual punishment and to define such punishment as torture (long periods of solitary confinement; confinement in cabinets barely large enough for a stool; confinement in "hell holes" exposed to the desert sun). Is there a chance that we might proscribe such diabolical means of torture? Fortunately, there is. No one uses the rack or the thumb screw to torture prisoners any more. Nor do we burn heretics at the stake, or boil them in oil, send them to the guillotine or to the axe-wielding executioner. Gradually, we have also come to agree that electrocution may belong among the diabolical methods properly defined as cruel and unusual punishment.

Abandoning the Death Sentence

Given that the primary ethical rule of an immortalist culture must be "thou shalt not kill," we would eventually give up execution by legal injection, execution by hanging, by poison gas, by firing squad, by *coup de grâce*. We would therefore confine a politician who swore that he would never make war and who then betrayed us to a reasonable number of years in a civilized prison, one without cells and with plenty of diversion and good food. Always provided, of course, that we can depose him without firing a shot. It really should not be too difficult to restrain such would-be tyrants. We could, for example, mandate the election of a praetorian guard responsible solely to Congress or to a parliament whose duty it would be to seize a warmonger and to deliver him or her to the criminal justice system.

Electing a Praetorian Guard

What would prevent a member of the praetorian guard from falsifying a bill of particulars against the Chief of State so that he himself might succeed to that exalted post? Well, a law might be written barring a praetorian guardsman from ever becoming a Chief of State by requiring that all Chiefs of State be freely elected. But what if the scoundrel yearned for no more than the hidden glory of being the power behind the throne and was supported by a third party not immediately recognized as the scoundrel's puppet? Then the people would have to rely on an independent police power to redress their grievances. It is easy to see why a system of checks and balances, not dissimilar to that provided by the US Constitution, would be imperative in a transitional society moving ultimately to a law-free society in which ethical behavior would be inbred.

The Futility of Utopias in Death Cultures

We have dreamed of such utopias more than once in the past but never have we dreamed of them in the context of an immortalist culture. It may be that ethical behavior of the highest order is impossible to attain in a death culture. A death culture teaches that death is inevitable after eighty, ninety, one hundred years. We are constantly reminded that we are destined to die — from cancer, heart failure, Alzheimer's or some rare disease of age (or youth) — so what is the point of sacrificing the certain pleasures of youth to sneak unobtrusively into the uncertainties and infirmities of old age? Wine, women, men, song, wealth, the sybaritic pleasures of youth — we want them all. Let no one stand in the way of our attainment of huge stores of wealth and the most handsome men and women money and/or fame can buy. What is the point of living to one hundred and three to witness the loss of sexual power and perhaps the very ability to walk and run? Some of us would gladly trade whatever it is that youth promises for the hope of reaching one hundred years. But the most aggressive among us scoop up whatever we can while the fires of life burn brightly, tossing all caution to the winds to have it all before death claims us. We thereby set the example which attracts those less talented but equally willing to sacrifice life, limb and liberty for their share of wealth and the fairest maidens. How, when they lack the talent to acquire great riches and fame? How else but by stealing, killing if necessary and, at the least talented but most aggressive level, killing for profit?

It is too much to hope that the lure of wealth, fame and high sexual success will be dimmed by the prospect of immortality so for a time, perhaps longer than we might wish for, there will be need to imprison street criminals as well as white collar and political criminals. Mercy would urge us to forgive criminals once, twice, perhaps three times and return them to society, presumably reconstructed. But in an immortalist culture it might be too much to demand understanding and compassion for miscreants who would view a thousand- or ten-thousand-year lifetime with contempt. What more on God's earth can there be to induce parents to teach their children not to maim, steal, kidnap, rape and kill than the promise of a long enough life to acquire all the money and all the mates one could possibly use and enjoy? Nothing. We might show mercy in our current death cultures simply because life is short and confinement for life too severe for all but the most recalcitrant killers; but in an immortalist culture we would have no choice but to punish without mercy anyone who would deprive a fellow human being of biological immortality. To steal away the life of a school-age child is horrendous enough but to steal it forever is to sin against all the gods who ever existed.

An International Prison on Antarctica

Isolation of criminals in island prisons comes with a distasteful history. Since the ethical objective in a transitional immortalist society is to imprint the caveat against killing, such societies would have little choice but to separate killers from the world in, say, a remote continent like Antarctica where they would be maintained without the blessing of telomerase replacement until they, the killers, died. An immortalist society would hardly owe a killer of someone who might have lived for 25,000 years the right to live that long himself or herself. Justice is fairness.

We have to assume that, like the turn of an invisible screw, the benefits of immortality, however relative, would beckon even the most determined sociopath. Potential criminals are nothing if not "percentage players." If a money-hungry sociopath earning thirty thousand dollars a year is tempted to embezzle a million tax-free dollars in exchange for ten or twenty years in prison, he is likely to yield to the temptation. One constantly hears of criminals calculating the percentages of crime and punishment. In an immortalist society the odds would favor crime even more than in our death cultures. What is twenty years in prison, even on Antarctica, when one has thousands of years to live? Legislators will have to find a commensurate punishment, perhaps, say, one hundred years for grand larceny, fifty years for assault and petty theft, three hundred years for rape and kidnap, and so on and so forth. Whatever terms legislators might agree would be most likely to change the odds in favor of non-criminal behavior; then, short of execution, let them impose those terms.

The Problem of Adultery

There are temptations as great or greater than money and the junk it will buy and one of those is the temptation of love or lust for a beautiful woman or a godlike man. Few events try a man's or woman's self-control more than the infidelity of a partner. Most often it leads to divorce (lawyers agree that adultery is the cause of 90% of all divorces in the United States) but more often than other motives, it leads to murder. Love and lust are difficult emotions to control. In our death cultures, where the motive is always to squeeze in as much fulfillment as seventy or eighty years will allow, caution is tossed to the wind like a shuttlecock, but in an immortalist society adultery may not be worth the risk of inflaming a jealous lover. There are saner solutions to the problems of love and lust than to challenge a man's or woman's pride, and we might well pursue them.

After all, if ten wives can live in harmony in a polygamous family (and in harems) and if one Amerind woman can make ten braves jump through her polyandrous hoop, it should be possible for people who will have conquered the problem of death to devise customs that will permit philandering without loss of honor by anyone. Rumor has it that an unnamed town in Northern Italy has adopted a custom whereby a married woman has the right to spend Wednesday afternoons with her lover. Any Wednesday. Being Italians, one can assume that by now the cuckolded husbands have learned who the cuckolded wives are and put the scales in balance. Italy begged for a re-customization of sexual mores. In Italy, to have been cuckolded was once a fate worse than the loss of a limb.

Certainly, voluntary or permissive adultery would be preferable to polygamy. The wonder is that polygamous marriage did not in its Mormon heyday trigger more violence than it did. In nature, the ratio of female births to males is 1.05 to 1. Not enough to justify harem marriages. What happens to the excluded males in a society in which one man is married to ten women? Do the womanless males, like the Canadian red elk, develop sexual strategies which allow one young buck to engage a dominant male in combat while another young buck mounts his wives? A dominant male walrus so weighed down with blubber he can hardly move is not so accommodating. He is quick to flop from one point to another within his harem to take on challengers in bloody combat. Polygamy is not a strategy for human beings blessed with extraordinarily long lives.

When Crime Will Disappear

What to do about criminal challenges in an immortalist society will be a problem confined to the transitional period — the time between the announcement of a series of successful telomerase reconstitutions and the day when we truly believe that we have cast off the curse of "natural death," which will probably be the day when a significant number of people celebrate their two hundredth birthdays in good health. The crime rate should fall precipitously on that date if it has not already have fallen dramatically. When the odds are such that no crime can justify the punishment, then crime will disappear and we can go about enjoying our endless days without fear of losing them to some maniac consumed by the need for vengeance for a real or imagined injustice or, what is more likely, consumed by self-hatred turned outward because he or she is deformed or is told that he or she is ugly and therefore unworthy of the most beautiful women or the handsomest men.

The Special Problem of the Genetically Deformed

What society will do with its genetically deformed is really not the problem it is in our death cultures where the compulsion to reproduce leads to dwarfs mating with dwarfs, thus sustaining the condition, the isolation and the humiliation for generations *ad infinitum.* In an immortalist society the compulsion to reproduce may be abated (we have presumed) and a model established that the deformed can follow without dishonor. Charity should remind us that deformity is not always the result of an unfortunate mutation. Just as often it is the result of an accident. On a sunny, fun-filled afternoon a godlike movie star is thrown from his horse and becomes a quadriplegic. It could have been worse; he might have been killed. Instead, he became an irrepressible lobbyist for intensified research into nerve regeneration which automatically made him a lobbyist for gene replacement therapy and an equally devoted promoter of stem cell therapy. So he served us all.

The Problem of Derring-Do

What to do about the accidental snuffing of a potentially immortal life by accident is a conundrum. Accidents like our horseman's, undertaken as challenges, may, like crime, also wither away. Taking six-foot jumps on a 17-hand stallion may seem like child's play to the skilled horseman compared to auto racing, parachute jumping, bungee cord jumping and other feats of derring-do, but these are macho feats intended to poke death in the eye and, when successful, are meant to win for the hero the applause and rewards of the crowd for demonstrating that death does not have an absolute claim on us. For where the odds say we must die, we thumb our noses at that desiccated monster and show that sometimes, nay, most of the time, we can conquer where we should fail — until the day when a seasoned tightrope walker falls to his death; a man shot out of a cannon lands on his neck; a mountain climber freezes to death at the top of Everest, and so on and so forth. Such death-defying exploits will have little or no meaning in an immortalist culture except as the folly of maniacs thinking they might impress women; or for women, the mania to prove they are the equal of men. At what? Defiance of death?

The Need for Maximum Safety in Travel

Still, there are the problems of automobile and airplane accidents and the questions of whether or not to regulate airline, train and auto safety even more rigidly than we now do. Will more and more regulation really prevent accidents? Apparently it can. In the generation and more that presidents of the United States have been flying in airliners reserved for presidential use, none has ever crashed. Now, such remarkable success may be credited to infrequency of use or to flawless maintenance. Whatever the reason, we might emulate it for public flights, which is to say: treat all citizens as we treat presidents and kings. Who is to blame when there is an airline accident (usually fatal for all involved)? When the fault is pilot error, the point is moot. But where the culpability lies with the builder of the airliner, the airline's mechanics, or the air traffic controller, then punishment is possible.

Accountability for Travel Accidents

To argue that no one would build airplanes (or repair them, or guide them in the air) if there were a chance that he or she would end up in Antarctica is to serve up another red herring. If doctors and surgeons can be held accountable for their mistakes, why not an airline-, a locomotive- or bus-mechanic? Indeed, locomotive engineers have been imprisoned when they were found to have been under the influence of marijuana or some other controlled substance when the trains they were guiding jumped the tracks. Why have airline mechanics, very often maliciously set against management, been treated as sacred cows? In our death cultures, people may shrug their shoulders in the face of a rash of accidents involving public carriers, assuming that death is the ever-present and unavoidable menace. But in immortalist cultures where we might enjoy limitless life, the response may not be so sanguine.

Why the Need to Hurry?

Will flying become obsolete? It is difficult to predict. The critical marketing questions would be: (a) Why would people be in a hurry? What urgency would compel them to run the risk of death in near super-sonic travel? Is an aging loved one approaching the end of his or her mortal life? Would we need to be there to bid him or her a fond and final farewell? Does business competition urgently require that we meet some monthly sales quota before the clock runs out? Second question: (b) Would people suffer the inconvenience of train, ocean liner and automobile travel? A tentative answer would be "yes," they would suffer the inconveniences of slower travel if trains were made to be as attractive as ocean liners and automobile travel were served by a motel and hotel industry that knew how to provide something more than a bed, a restaurant and, occasionally, a swimming pool. The Marriott hotel chain also provides (here and there) exercise gyms. Las Vegas makes luxury hotels lures for gamblers. Given that for immortalists boredom will weigh as heavily as illness does for us mortals now, we can expect entertainment to be the dominant industry of the future, but the odds are that it will be as regional as it is nationalistic now, and for similar reasons: language, travel, cultural differences. (Note the spectacular welcome given to a country singer performing in New York's Central Park. Note also that he seemed to require the auspices of a native New Yorker.)

Immortality Will Not Justify Indolence

The pervasive deceleration of activity in an immortalist society, especially behavior entailing a high risk of death, reinforces the idyllic notion that because we would be immortal we would have no need to work. Since we would be immortal, how could we possibly die of starvation? We wouldn't. There would always be soup kitchens for the indolent. But the need to work, if not the anxious, overly-competitive reasons why we work now, would be almost as demanding. We work now to provide food and shelter for ourselves and our families. We also work now to put something by for a rainy day, which often translates into a college fund for our children and retirement for ourselves (retirement meaning when we are too old or infirm to work). In an immortalist society we might not work to support families (which would be in a pattern of rapid obsolescence) but we would certainly need to work to provide food and shelter for ourselves. We would soon discover that unless a majority of the population worked, our supermarket shelves would be empty of produce, canned goods, meat, fish, soap powder; there would be no new cars or trucks to replace used cars; there would be none of the products that make living easier as well as possible. Medicines, toilet paper, pots and pans, the very medical treatments that would spur us into each succeeding phase of our long extended lives. Would we revert to primitivism simply because we were no longer racing against death? Not likely. We might not work competitively, vying with each other for recognition, but we would work not to lapse into life-threatening privation.

The Replacement of Death with Boredom

We would also work to relieve the cloying problem of boredom. Boredom will replace death as the universal nemesis. How many books can one read, how many pictures can one paint, how many sculptures sculpt before *ennui* overtakes us with its sickening nausea? Will it be possible to play video games, or chess, or basketball and tennis all day long, year after year after year? Yes, for champions who cannot step out of the limelight, but even great tennis champions take up golf and great golf champions take up basket-weaving. Champions are made of more single-minded stuff than the rest of us. The glory of being a champion seems to outweigh the repetitiousness of the game by an order of magnitude.

Inevitably, given the enormous time spans of an immortal lifetime, we will all be rich. We may all be willing to return to work if for no other reason than to relieve the onerousness of boredom. But will rich people cook in restaurants? Will they operate locomotives, repair cars, grow food, milk cows, sweep sidewalks, catch and prosecute criminals, make and interpret law, dispense medicine, continue medical research, build space ships, sew clothes, mine coal, drive buses, operate ocean liners, make movies, write and publish books and magazines? They will. Boredom and the evidence of empty stores and gas stations will drive us out of retirement. We will be delighted to cook in restaurants, to operate ocean liners; to fish for shrimp and tuna, to run farm machinery and repair automobiles even with millions in the bank. Being rich and idle works only when there is a worker class to support the sybarite.

Upgrading Incentives for Degrading Labor

One might argue that it would be all well and good to return repeatedly to one's basic career or to learn new occupations, but who would return to collecting garbage, cleaning cesspools and public toilets, handling other public waste, slaughtering animals, sweeping streets, mopping floors in office buildings, airliners, trains, court houses, and legislative halls? In our death-doomed societies there seems to be an almost endless supply, generation after generation, of the poor and untalented ready to accept whatever vile occupation the rest of us will not. They are paid the most humiliating wages, are usually without union representation, and are thus unable to put aside for a rainy day or to accumulate a nest egg to bequeath to sons and daughters so that they, in their turn, might avoid the ignominy of base labor.

In an immortalist culture with no pressure to reproduce oneself and with all the time in the world to amass small fortunes, we may well run out of laborers sooner than might be expected. Even now they are learning to gain some dignity by demanding higher pay and retirement benefits. With these emoluments in place it should not be long before the cleaner-uppers of our daily messes will retire and stay retired; or if indeed boredom drives them back to work, they are not likely to return to the mortifying labor of mopping floors and cleaning toilets. What then? Inordinately high pay or a mock parody of Mao's Cultural Revolution during which lawyers were drafted to work in rice fields? Not likely, in the pampered, elitist West. Instead, we might expect a reversal of the indignity our servants now suffer. To entice people to perform menial and nasty tasks, we might give them the sole right to live in penthouses, drive luxury cars, and enjoy free tickets to concerts, plays and operas.

Who Can Afford Social Security?

There are larger problems with the work ethic than who will perform degrading tasks. We have observed that with the obsolescence of families it will not be long before everyone has accumulated enough money to retire. Say, two hundred years. Until then, millions of people may feel that they have earned the right to retirement on Social Security income. But when millions of people are scurrying to embrace the life of the lotus-eater and living interminably — who will finance government-sponsored Social Security? Theoretically, no one. As of this writing (2007) a base of 163 million American workers contributing nearly 10% of their gross wages can barely support 50 million retirees. How could a near-zero worker base support 300 million retirees? It couldn't. What then? Legislators and statisticians serving legislators feast on such problems. As the balance between the retiree group and the worker group teeters in favor of the retirees, we can expect statisticians to alert Congress (or parliament) to the imminent need for a change in Social Security law. In America, Congress has not always perfunctorily ratcheted up the percentage of a worker's pay that must be paid into the Social Security Fund to meet a crisis in funding. Congress has also enacted legislation permitting 401K, IRA and other self-managed tax-free programs which would allow high-wage earners to amass small fortunes in the space of a single lifetime. But all nest eggs, including such tax-motivated funds, depend on dividends and interest for income else the capital would soon be depleted. Before we become aware that most of us will have to return to work, or continue working, who will produce the goods whose sale will yield the profit which will, in turn, generate the dividends and interest expected on vested funds?

A Progressive Tax on the Rich

One solution would be a progressive tax on the rich. How easy it is to pick on the rich. Truth is, the incomes of the rich and the higher wage earner have been spared the full burden of payment into the Social Security Fund. The law has always limited the amount of FICA (the contribution to Social Security from wages) which high-wage earners need to pay. Currently, high earner or low, each need pay no more than 10% of the first $20,000 of income. (It started out as 2% of the first $5,000 in the mid-1930s.) If you're a coal miner earning $30,000 a year, you pay the same amount of FICA as the fellow earning $300,000. But the rich and the well-paid also collect the maximum benefits since benefits are awarded on an average of one's best and one's lowest earnings years.

(There is a movement afoot to get rich retirees to give up their Social Security benefits. The super-rich make a show of giving as much as 10% of their incomes to charity. It is not a show with wealthy individuals like Ted Turner, who once pledged the equivalent of one third of his assets to the UN to be used at the UN's discretion.)

As if the special exclusion from Social Security taxes weren't enough, during the Greedy Eighties the then-president asked Congress to permit the federal government to borrow the monies in the Social Security Fund by incorporating the Fund within the General Budget (where it still is). The rationale? Why pay foreigners interest on the national debt when the government could borrow from the Fund, without having to pay interest but pretending that it would if it became necessary? After all, reasoned the president, the government was ultimately responsible for Social Security payments, was it not? The whole truth was that transferring the hundreds of billions in the Social Security Fund made the Federal Budget deficit seem smaller, which is probably why Congress went along with the scheme. Why didn't (doesn't) it make sense to pay one's own citizens the interest on funded debt (bonds) rather than pay it to foreigners? It does — if only the government would pay it. The whole truth (again) is that during the Greedy Eighties this same Republican president and this same Democratic Congress agreed to finance budget deficits by selling long-term bonds to foreigners (principally England, Germany and Japan) rather than tax the rich. The result was that at the end of this vile period of betrayals the national debt had soared from $1.5 billion to $4.5 billion and the interest on the debt increased to $350 billion annually. (Be patient, there is more.)

One would think that with the restoration of the Republican presidency, this time with the support of a Republican Congress, a nation shamed by the deceits played on its citizens would use a burgeoning budget surplus to shift the Social Security Fund "off-budget" and return it to its sacrosanct independence fully reimbursed for twenty-odd years of lapsed interest payments. Heaven forbid. What the new president promised in his election campaign, and what he won, was a rebate of taxes which he freely confessed would be a windfall for the rich. They're the ones who pay the taxes, aren't they, he asked? It's only fair, he whined, that they should receive the largest share of the rebate.

A Retroactive Tax on the Rich and High-Wage Earners

It would have been more fair to require the rich and the high-wage earners of the eighties and nineties to re-pay the interest the nation incurred on its foreign debt during those years, debt incurred primarily because they were not taxed sufficiently to pay for increased national defense spending. An average of $250 billion a year in interest payments on the post-Vietnam debt multiplied by 20 years comes to $4.5 trillion in back taxes owed by the rich and the well-paid from the eighties, nineties and the '00 years. Purely by coincidence, $4.5 billion is a close approximation of the funded national debt. A special progressive tax of, say, 1% or 2% on assets over $1 million would easily erase the debt and with it the $400 billion in annual interest charges now paid to foreigners. Well, that's a bit demagogic. Some of our long term bonds are held by Americans. Saving $400 million in annual interest payments would inevitably allow a reduction in income taxes. (It is really all moot. When confronted by special taxes, high-wage earners merely persuade their Boards of Directors to increase the number of stock options granted to them for the following year. The highest wage earner among CEOs in the year 2007 was paid $158 million, mostly in profits on stock options. Playground people would call that a slam dunk, which is not dissimilar to shooting ducks in a barrel or fishing for eels in a tank).

A Progressive Tax That Would Wipe Out the National Debt Would Be a Wash

Getting Congress and the president to tax Americans for avoidance of taxes in the Eighties and Nineties has about as much chance of realization as a snake has of catching moonbeams to make a pearl necklace. But someone has to pay that debt. The sniggering hope of the high-wage earner and the born millionaire is to shift the burden to future generations. They won't specifically say shift it to their sons, daughters and grandchildren, but that's what they mean. Bean counters to the rich might remind them that the interest earned on great piles of wealth even for one generation would go a long way toward defraying whatever they might be taxed to liquidate the national debt. Two or three generations of unearned interest income should easily complete the liquidation.

One's heirs and heiresses should not complain if the piper comes acalling in the next forty or fifty years. The fondest hope of the rich, of course, is that it will be your children who will pay off the debt, not theirs. Next to that is the cherished hope that the national debt will somehow go away; become worthless like the Liberty Bonds of World War I or the Whoops bonds issued by the State of Washington during the 1970s and 1980s to finance its atomic energy for electricity program. It is not a likely scenario; the rich and their bean counter puppets know it, so the game they play is the one all scoundrels play: wait for inflation to allow payment of the debt in the cheaper and cheaper dollars of the future. Normally, paying off dear dollars with cheap dollars is the game played by borrowers, which is why, if you don't already know it, the rich panic in the face of spiraling inflation. For the rich, runaway inflation is a perilous game, as the memories of marks being pushed around in wheel barrows in post-World War I Germany will attest. How much more sense it would make to pay off the national debt and thereby reinforce the world's confidence in the full faith and credit of the United States of America. Would it not be a glorious day in the history of the world for common sense and charity to overcome greed? What a marvelous day it would be to watch a million millionaires behave like Ted Turner, who quickly saw that having three billion dollars was a self-mocking absurdity. What on God's good green earth does one do with three billion dollars, which at 10% earns $300 million a year or close to a million dollars a day? What Turner did with it: donated $100 million a year to the UN and then picked up the difference between what the UN said the USA owed in delinquent dues and what Congress was willing to pay.

The Threat to the Oligarchy

Congress, however, is a flock of sheep largely directed by the rich who finance their campaigns, and some among the rich cannot bear the concept of a United Nations because it threatened and still threatens the oligarchy which the United States was becoming and has become. Bill Clinton's determination to reverse the oligarchic process was the principal reason he was so fiercely hated by the rich, who sowed their hatred wherever they thought it would bear fruit: mainly in the Plains States and in the South. How? By attacking "Big Government," which translates into high taxes, racist preferences, and support for the poor. It is the profound wish of the rich and the sheep who fertilize their fields that the federal government will one day rescind personal income taxes and the tax on estates and allow the cost of government to be defrayed by sales taxes, which is in effect a tax on the poor and the low-income wage earner. The reasoning is simple, if not simpleminded. Why shouldn't everyone share in the costs of government; why should the rich bear the burden alone? Could it be because it is the poor and the lower middle class who fight their wars and work their mines and risk their lives as policemen and fire fighters? Ted Turner may not save the world (the world will be saved by microbiologists), but his gesture may well be the kind that could save American democracy.

Like It or Not, The Rich Will Share Immortality with the World

Who knows? Perhaps the prospect of living thousands of years will dissipate the iron tenacity of greed. What meaning will billions of dollars in personal wealth have when the most precious of all rights — the continuous maintenance of life — is available to all? The rich who resist a national health program (currently enjoyed by all people in the industrialized world, except in the United States) had best gird their loins and dig in their heels against the day when every citizen of the world, not just the US, will be entitled to stem cell or gene replacement therapy. Can anyone still in possession of his and her faculties imagine a world in which the citizens of one nation enjoy relative immortality while the rest of the people in the world continue to die? Impossible. Even the rich will not deny the world that blessing. After all, what can it cost? Stock market speculators may rub their hands with glee at the possibility that one publicly traded biotechnology company will own a patent on the procedure that will guarantee relative immortality. Will the sky be too high a price to pay for telomerase? Actually, it probably will not matter. Countries like Sweden and Germany will probably make the procedure available to all without charge once the inventors are properly rewarded and the costly years of unsupported research and development are paid a hundred fold. Will ten billion dollars reward the inventors and the pioneering investors? Let the federal governments of the world pay it so that they might immediately prevent the sort of stock speculation that puts the price of less spectacular drugs beyond the reach of all but a fortunate minority. Some things are best administered by governments: war, policing crime, fighting fires, prevention of epidemics, and soon the prevention of what we euphemistically call "natural death." Like the twin vaccines for prevention of polio in the mid-twentieth century, telomerase will be distributed freely or at token cost to everyone in the world.

A Tentative Blueprint for the Future

One day soon, we will all be immortal; or our children and grandchildren will be and they will adopt an unqualified rule against killing. Nations will denounce war and revolutionaries will forswear violent revolutions. Or else. Or else what? Will we bomb them into submission? Will we caution them not to kill or be killed? We have already been through this vicious claptrap and once again, for the umpteenth time in history. By now, *surely* we have learned that war and bombardment reap nothing but mass graves and murderous grievances stored against some future day of retaliation. Haven't we?

There are non-violent ways to encourage compliance with the rest of the world. There is ostracism from the family of nations including economic and scientific embargoes which will, after a century or two, encourage compliance. There may or may not be the familiar figure of the international smuggler measuring himself against the power of nations to violate embargoes and other sanctions. So vital is it that all nations agree to give up war and revolution that convicted smugglers breaking embargoes should be treated as killers (they would clearly be aiders and abettors of mass murder) and sentenced to Antarctica, where they will be allowed to live out their lives without the right or the means of extending them. Mind you, the purpose of confinement on Antarctica would not be rehabilitation but punishment of sociopaths who would callously trade millions of human lives for millions of dollars. And it must be done not by killing but by attrition and by the example which attrition sets. There cannot be any violation of the preeminent rule of an immortalist culture, which is the prohibition against killing by individuals or by the state. If there were no other rule or law in a society of immortals than this one, it would probably be all the law we would need. Cull from existing and future generations devils who would equate millions of dollars with endless life and the world will soon find itself in the condition of peace on earth and goodwill toward men which it has dearly desired — but which it has not and will not achieve so long as all believe that however we try to master it, life is futile; that we live on the edge of the abyss; that if we don't die in one hundred or two hundred years, we will expire with the death of the universe.

The Universal Human Goal

What do we do about philosophers, minor pundits, and cultists warning us against the folly of believing in immortality by reminding us that the universe is ultimately doomed some five, ten or twenty billion years from now? Pay them no mind. Or ostracize them. The principles of free speech and a free press are worth all the venom the fangs of the naysayers can inject into the mainstream of an immortalist society. Ostracism is a noble substitute for guns and torture and should be used freely and liberally against criminal nations, obstructionist politicians and journalists. Do not trade with criminal nations; do not vote for obstructionist politicians; do not buy the books, magazines and newspapers they are allowed to publish. Never be dissuaded from the universal human goal: the conquest of death.

An immortalist society cannot long suffer the widespread ownership of guns. It is imperative that ordinary citizens be prohibited from owning guns. We have seen the high resistance gunsmiths and their lobbies can mount against gun control. The best strategy would be to ban the manufacture of guns; then, as the possibility of relative immortality begins to seep into the collective consciousness, ban the manufacture and importation of ammunition and, eventually, confiscate the existing supply of guns except those required by the police and the armed forces. It may take fifty or one hundred years, but if we are purposeful, it will be done. In the meantime all violations of gun control laws should be treated as felonies carrying life sentences on Antarctica without hope of parole and without hope of telomerase replacement.

The problem of how to persuade all nations to agree to help build one gigantic international prison on Antarctica and then agree to store all of their criminals there is a matter best resolved by diplomats. There may be some who resist the idea of imprisonment in so forbidding a climate as cruel and unusual punishment. But it is the very climate and isolation of that vast continent (10% of the world's land mass) which would make escape impractical, if not impossible. Odious comparisons will be made with Devil's Island and Alcatraz, which eventually were proved to be not so restrictive as we believed. It remains to be seen how the attitudes of terrorists and gangsters will change, given the prospect of relative immortality. Always we shall require the prism of immortality in our telescopes when attempting to compare historical behavior of death-doomed people with the future behavior of people who know they are death-free.

The prohibition against killing by the individual or by the state leaves little choice but to ban abortions except in cases of rape, incest or deformity of the fetus. We need only to be reminded that each of us has been immortal until now (the sex cells which generate us have never died; each preceding generation of parents had to have been alive when it conceived the next generation) to see that abortion presents a grievous problem. The state should provide a sanctuary where young women may find anonymity during the late stages of their pregnancy and where barren women may find a child to love and rear. While state support for adoption of unwanted children would seem to run counter to the probability that human reproduction will obsolesce in an immortalist culture, it isn't, or needn't be.

Just as we now fear the prospect of overpopulation, so too must we fear the prospect of under population. Accidents and epidemics will take their toll and over long periods of time will endanger our ability to provide water, power, food and services (ambulance, police, health care) critical to the maintenance of life. As noted, one of the great ironies of an immortalist future may be that we will have solved the problem of death but will still drop like flies in the grip of an unexpected epidemic (AIDS, Ebola, bird flu are the latest examples).

A Committee to Appoint the Supremes

That it is the right of a woman to abort an unwanted pregnancy is the misguided judgment of the US Supreme Court, which, in the name of the various rights guaranteed by the Constitution, can effectively make law. Since the Supreme Court reflects the liberal or conservative mood of the country at any given time, it would be advisable to amend the Constitution as soon as possible to take the power to appoint justices to the Court away from the president and give it to a committee of non-partisan deans of law schools, police chiefs, doctors and clerics. The instruction to such a committee would be constantly to review the careers of lawyers who display a non-partisan commitment to the interpretation and application of the laws that the people, represented by Congress, would enact.

The Praetorian Guard

The most ticklish change an immortalist would undertake in the American or in any other democratic system is a constitutional amendment empowering a so-called praetorian guard, for want of a more appropriate name. As noted, the primary mission of such an institution would be to check and balance the excesses of a president who pledges to denounce war and then surreptitiously leads the nation into one. But the power of such an institution could include a mission to prevent dictatorships whether of the right or the left, of the executive, legislative, judicial or military variety. The unit should be a military force large enough to deter a *coup d'état* by the principal armed forces of the nation but small enough not to be itself a threat. In a world where we have vowed never again to make war, it would be an ideal way to re-define the mission of the US Marine Corps. The Corps, as it likes to call itself, has a two hundred year history of independence, bravery, achievement and loyalty to the American constitutional system. And it comes with a ready-made credibility. In a minor way, the Corps now performs one of the functions of a praetorian unit: it guards our embassies everywhere in the world. Taken out from under the thumb of the military, the Corps as a praetorian unit would be responsible directly to the people (Congress, parliaments).

In light of the confusion surrounding the federal election of the year 2000, once a law establishing uniform voting procedures throughout the states is enacted a praetorian corps might take on the supervision of federal elections, removing that right from each state where it now rests. The decision by the Supreme Court which effectively chose the winner in that election was a cynical violation of the separation of powers. How would the Supreme Court have acted had there been a praetorian guard? It would not have acted at all. The praetorians would have been in charge of the election and would probably have opted to call for a new election. We trust that the establishment of uniform voting procedures will obviate such action but, always guarding against the unseen, we need to be prepared to prevent dictatorship by any branch of federal government. In the recent past the danger of tyranny came from the left; recent elections have reminded us that there is also a danger of tyranny from the right, and not only from fear of usurpation or silent overthrow of a constitutional society but for the crass purpose of an income tax rebate and the more insidious ambition of institutionalizing an oligarchic society. How is that ignoble goal pursued? By working toward the elimination of income and estate taxes. By establishment of a national sales tax. By using bonded debt instead of taxes to defray the cost of government. By encroachment on the environment ostensibly to provide energy but actually to create one more get-rich opportunity for now obscenely rich corporations and individuals.

How does the entrenchment of an oligarchy threaten the aims of an immortalist society? By granting to a money-crazed elite the power to decide whether or not it is the obligation of the federal government to extend the blessings of long-extended life to the less fortunate among us. The right to enjoy relative immortality must be an absolute right, non-negotiable and unimpinged upon. Given the possibility of living hundreds, even thousands of years, we will one day all be rich. Let us pray, nay, let us make certain that we will be led by armies of anti-oligarchists like Mr. T. and not by money suckers who can find no more noble purpose in life than the accumulation of more and more and more and more and more and more money.

Chapter 8. The Three Compulsions (Contributions to a Theory of Quantum Psychology)

Love of money is not the root of all evil; death is. We are born innocent and the world teaches us to believe that we are surrounded by things, people, and animals who would do us harm. We must therefore learn to defend ourselves against all predators, human and otherwise. It doesn't take long to learn that our mothers and fathers have correctly evaluated the world and for us to steel ourselves against a manifestly hostile environment. The pussycat scratches when we tug at its tail. An older brother and sister will not allow us to play with their things and our mothers are forever telling us "no-no" and occasionally slapping us on the wrist to reinforce the message. What on earth is going on?

To begin with, we are born innocent, innocent that is, of death and the fear of death. The reproductive cells out of which we were formed never experienced death. Our mothers and fathers were both alive when they conceived us as were their mothers and fathers when they were conceived. As far back as we are willing to go, our reproductive cells were free of the stain of what we call natural death. We don't learn about natural death until an older neighbor crushes a caterpillar, or a fly, or chops the head off a sweet water bass while he matter of factly teaches us that we don't like to eat fish heads. Anyway, that's how we kill them. Kill them? Good Lord. Do we have to kill fish in order to eat them? Yes, we do. And do they never come back? No, but the Mama and the Papa fish make more fishes. But not the same fish, no, those fish are dead and gone. They went into our stomachs to help us grow big and strong.

One day you learn that there are people who eat nothing but plants, which sounds great until you learn that they don't kill any animals at all, including man-eating tigers and snakes. You learn that you have to be careful when you go to India not to kill cobras, even though cobras kill 20,000 Indians every year. To hell with that, you say — who wants to go to India? Your friend advises you not to make a big deal of it; you're going to die anyway. Who me ? Chop my head off? No, nobody's going to chop your head off. You just get old and you die. How old? Depends on the person. Some people live well past a hundred years. And some people die when they're six years old. Six years old? How can you die when you're six years old ? Easy. Cancer, starvation, malaria, drowning, all kinds of ways. Not me, buster, not me. Somebody else but not me.

At this point in a child's development, something clicks inside the brain. The brain reasons that since kids can die when they're six years, even younger, you're going to be sure you're not one of them. One way to be sure is to defend yourself against all predators, which includes everything and everybody else trying to escape the death trap; and to make sure somebody else dies instead of you — not by murdering someone but by avoiding the draft when there's a war. And never riding in the passenger seat of a car.

Certainly not by swimming across the English Channel. Or joining the cops or firefighters. Or building skyscrapers. No, sir. Call me Dance-Away Harry. Let somebody else do it

Look, kid, says an older, street-wise fellow. You can't escape dying. Someday you're going to get caught. A fire in a circus tent A hurricane. Two guys robbing a bank end up shooting at the police and you catch one of the bullets. If that doesn't get you, old age will and you'll die in your sleep. You can die your sleep. Die in my sleep!? Sure, kid, you can die lots of ways. All you can do is be smarter than the next guy and don't slip on a banana peel or do a back flip in a swimming pool. And don't take any wooden nickels.

Greed, says an old lady, greed is what you gotta watch out for. It starts out all right. You collect stuff to make your life easier; then it gets you. There isn't anything you can do in moderation when it comes to money, power, women, wine and art collections. So relax, you're not going to go down on the Lusitania and you're probably not going to die in your sleep. You're going to be immortal, so learn about the three compulsions and don't worry about greed. Pretty soon everyone will believe that he and she will not die a natural death and everything and everybody will slow down a bit.

1. The Compulsion to Reproduce

The compulsion to reproduce would cover all behavior related to sexual behavior: mating to reproduce; mating for the pleasure of mating with no thought of reproduction; and lust, which we would probably define as an insatiable sexual appetite but which should be placed under the heading of greed. Greed is the endless accumulation not just of money but of everything. (Note: By "compulsion" I mean unwilled or unconscious behavior by the organism, any organism, much like the behavior of our autonomic nervous systems.)

2. The Compulsion to Maintain Form

The compulsion to maintain form is almost as obvious as the compulsion to reproduce; all it signifies is the automatic behavior to resist disintegration or deformation — what we mean by death of the organism. The belief that death is inevitable motivates inquisitiveness and accelerates ambition. Greed is thus the handmaiden of death and the instigator of crime.

3. The Compulsion to Aggrandize

The compulsion to aggrandize describes the phenomenon in nature by which organisms grow larger and more complex in size. The simplest insight to the phenomenon is Darwin's belief that organic life began in the form of a single cell which thereafter "aggrandized" to become the spectacularly huge and enormously complex symphony of the species.

The aggrandizement of an organism would appear to be something that the atom itself does automatically; it tends to want to form aggregates with other atoms and stubbornly to hold on to such forms until a competing aggregate has forced its deformation. Or vice versa. The atoms thus victimized simply pick themselves up, dust themselves off, and start all over again. All this while billions and billions of galaxies — galaxies, mind you, not just suns — are flying outward in space at speeds approaching the speed of light

Is Greed the Ultimate Emotion?

Whether or not greed as an emotion is rooted in the compulsion to maintain form or the compulsion to aggrandize would be a debate for philosophers who like to count the number of angels who can dance on the head of a pin. Without the ability to maintain form, all other considerations are moot, so greed ultimately serves maintenance of form as it voraciously seeks aggrandizement. The virus, bacterium and fungus appear to be quite happy with their simple organic forms (each is immortal), though each is subject to mutation. Every day, it seems, we hear of new strains of bacteria and viruses.

Mutation is another way of understanding how the atom victimizes — or benefits — a particular organism. The accumulated benefits of mutation may take the form of sheer size (dinosaurs, whales, elephants, sharks) or they may take the form of a bigger brain or an opposable thumb, as it did with us.

While greed might, therefore, find nobility in the compulsion to aggrandize ("See, I can't help myself, it's the way the atom works"), aggrandizement at the expense of another organism deprives it of the ability to maintain form, thus debilitating or destroying the organism immediately, as a lion devours a zebra or as we slaughter domesticated animals.

We could reasonably argue that it is nature which sets up the competition making one organism dependent on the consumption of another in order to survive, but that argument may also be moot. Rather than the way mutation dictated aggrandizement, it may be the way the big cats adapted (huge jaws, dagger-like teeth, high speed). Rather than spend the entire day and night grazing as zebras do, cats might have learned early in their evolution that it is much more efficient to kill and eat another animal for the food they need and then spend the rest of the day, or days, sleeping or making love. We may sneer at the savagery of predators hunting prey in the wild but we rarely hear of predators killing more than they need to eat at the moment.

From what biologists and zoologists tell us, or show us on film, animals in the wild do not pile up carcasses as provision against next week's hunger pains. A big cat may drag home the chewed-over carcass of an antelope for its very young but not as goods to be saved until the next time hunger gnaws at his or her stomach linings. Apparently, wild animals prefer their meat fresh. (Apologies to the dog who has learned to bury bones, with or without meat left on, against future need; but then dogs are brighter than we think. Apologies also to squirrels and mice who pile up acorns and other nuts against winter.)

We cannot fairly accuse all wild animals of greed although, judging from nature films, it would appear that big cats will snarl at ambitious young cats who dare to think that they can dine high off the hog, side by side with the master, before he has had his fill. Monkeys and chimps aren't much better behaved. Jane Goodall showed us a mother chimp who, coming upon a basket of bananas conspicuously placed by Goodall, ignored the pleas of a hungry youngster until she'd had her fill. Yet the legendary African wild dog, a ferocious predator, will feed her young before she even thinks of taking a bite. And who better than all of our winged friends demonstrate the principle of children first?

The Greediest S.O.B. Outside of the Sea

Feeding oneself before feeding one's young could legitimately come under the heading of maintaining form but it may also come under the heading of greed, depending on other factors and other behaviors by that particular animal. The lion, say. Who hasn't seen the documentary in which a lion steals a carcass from the lioness who brought it down (daring her to challenge him)? In India, lions will eat the cubs sired by a preceding male or males. (In India, lions seem to travel in pairs to fulfill their reproductive urges; the lioness is apparently a tirelessly demanding lover until she is as certain as she can be that she has been impregnated.) Clearly, the lion and other big male cats are specimens of an overreaching compulsion to maintain form at the expense of others and therefore prime examples of raw and naked greed. The evidence indicting them is compounded by their apparent indifference to whether or not females or cubs are fed or not. On all counts, therefore, the big male cat would appear to be the greediest son of a bitch in nature (outside of the sea).

While this characterization of the big male cat may seem unfairly judgmental (after all, it is the accumulation of beneficial mutations which has made him a big cat rather than a tabby), it is not. Nature may have made him a bully but some large animals are gentle creatures; the whale, for example, who feeds on plankton; or the elephant who is an herbivore and most of the time is not a predatory killer. How could he be? He/she eats leaves. But we have seen news photos of circus elephants gone berserk who kill their trainers and documentaries showing bull elephants running amok in India. While size (aggrandizement due to accumulated mutations) is to be blamed or credited for the enormous appetites of elephants, sharks and big cats, size is not enough, by itself, to account for aggrandizement. Apparently, the tendency in large animals to be gentle or ferocious, "reasonable" or overreaching, is a matter of genetics, which is to say mutation and adaptation — but that is not likely.

Synthesis of Protein, Not the Result of Adaptation

The ability to synthesize protein from grass and leaves, which herbivores do, cannot be the result of adaptation. For the converse genetic reason, carnivores cannot synthesize protein from grass and leaves and must eat the animals who can. Does carnivorousness justify overreaching? No. There are too many examples of hyenas sharing a carcass and vultures sharing a carcass to argue that carnivorousness justifies overreaching. The theory that eating red meat, or any meat, is what accounts for aggression needs to be reviewed.

Overreaching Behavior Is Inborn

It would appear that overreaching behavior is mandated by mutation. Some large animals are highly aggressive, some rarely if ever aggressive. But aggressiveness is not limited to large animals. In his study of aggression (*On Aggression*), Konrad Lorenz concluded that the snake and the Norwegian brown rat are among the most aggressive of all species. (Let's not forget the pit bull terrier.) Lorenz once left a pair of love birds alone (in their cage) while he went off for the weekend, only to find upon his return that the male had almost totally destroyed his loving companion, leaving only the feathers. In *The Territorial Imperative*, Robert Ardrey, no mean anthropologist himself, tipped his hat to Lorenz, calling him father of the theory that aggression is innate.

Aggression appears to be inborn and not learned behavior and is probably the cause of over-reaching behavior that in humans is an expression of greed.

Must we now attribute a fourth category of behavior to mutation, namely, the compulsion to be aggressive? No, because aggression is not universal; indeed, if we judge by the number of species which are non-aggressive and those which are, the scales might balance in favor of a greater number of non-aggressors, denying the classic aphorism that nature is red in tooth and claw. The big fish eat the little fish, it is true, but elephants, moose, camels, alpacas, llamas, deer, elk, horses, mules, buffalo, hippos and giraffes eat leaves and grass.

Aggression Must Be Tamed

While the analysis of aggression is properly the academic domain of zoologists, anthropologists, and psychiatrists, and not philosophers, it would seem, on the surface at least, that (greed = overreaching = aggression = aggrandizement) is innate and must be tamed. In nature, the tendency has been to allow nature to run its course; in human nature, ethical religions have, over five thousand years (counting back to the early Egyptians and Sumerians), attempted to modify aggression and greed by making non-aggression the price of a paradisiacal life after death (Buddhism and Judaism reverently excepted).

The Ancient Egyptians Lived the Ethical Life

It would be comforting to observe that religions have succeeded in taming aggression and modifying greed, but history delivers a contrary judgment. Where our primitive ancestors fought for the possession of women, their civilized progenitors fought for territory, gold, silver and jewels, as well as women. Worse, the wars of civilized societies have often been fought in the name of religious truth, and in our century in the name of ethical ideologies. The major exception to this sanguinary rule is ancient Egypt in its pre-pharaonic phase (3000–1500 BC), which preached and practiced that the life lived ethically was sufficiently rewarding unto itself to seek no other motive. The result is a dearth of Egyptian law for this period in its history but an abundance of literature extolling the joys of the ethical life. The pre-pharaonic Egyptians also lived peacefully with their neighbors, though they were clearly the most powerful nation in the ancient Mediterranean world.

(The Judgment of Osiris religious system which made a life lived ethically the precondition of a joyful life after death did not imbed itself in Egyptian culture until the pharaonic age, 1500 BC and after.)

The remarkable ability of the early Egyptians to live ethical lives does not detract from the noble intent of the Golden Rule, the Ten Commandments, and the Sermon on the Mount to institutionalize ethical behavior in the Old World. Christians, attempting to live by Christ's self-abnegating Sermon on the Mount, suffered three hundred years of martyrdom only to usher in two millennia of religion by the sword which, not incidentally, included the Muslim conquests as well as the Christian. No one still possessed of his sanity would argue that post-Egyptian religion has succeeded in modifying ethical behavior or curbing the irrepressible call to arms by the most aggressive among us.

Taming the Lions and Eagles Among Us

While it behooves us to acknowledge that there are nations, such as Sweden and Switzerland, that have foresworn war, and nations like India that freed itself from its colonial masters by passive resistance, we need also to be reminded that Gandhi was assassinated by a disciple, that Jesus was betrayed by Judas, and that one of Gandhi's disciples, Martin Luther King, Jr., was also assassinated. Passive resistance was a successful political strategy in India (heaven only knows how many people would have been maimed or killed in an armed uprising) and a partially successful strategy for black Americans. It is sad to say, however, that in a world teeming with aggressors, passive resistance would merely inflame the appetites of would-be conquerors and un-Christ-like messiahs. Should we therefore first set out to tame the lions and the eagles among us? And how would we do that?

In an immortalist society, it would not be long before a literate majority recognized that military adventurism is a false pursuit which once served the equally false goal of historical immortality. In a culture which assumes the inevitability of death, the biological compulsion to be immortal has been translated into the need for historical immortality: actual history book mention, just as the would-be conqueror has seen other conquerors mentioned. Historical immortality would also include fame and glory: press mentions, interviews, statues, statuettes, recognition by head waiters and so on. But a majority, even when literate, is powerless to act except by mob rule, which ultimately leads to maiming, arrest, and death depending on whom the police and the armed forces support.

The New Lion Tamers: A Praetorian Guard

It is imperative, therefore, that one of the first institutions established by Congress or Parliament be a praetorian guard (see above). The prime duty of a praetorian guard in an immortalist society would be to arrest a would-be conqueror after his first speech about ethnic cleansing, or *lebensraum*, regional and national interests, and have him brought to trial. If found guilty of inciting revolution, war, or any use of force, he would be packed off and sent to Antarctica for the rest of his natural life. Too severe a punishment for one who would incite crimes against humanity? He might be given a second chance to see if the old siren songs of historical immortality, power, fame and glory have stopped nagging his consciousness; if he persists, well, what other choice would there be but to commit him to Antarctica? A good guess would be that such super-aggressors would continue to crop up during the transition period from our present death culture to an immortalist culture.

Where we might fail to institutionalize a praetorian guard, the fault would probably lie with politicians reluctant to give up power. In the United States, for example, Congress enjoys ultimate power. It can override a presidential veto by a two-thirds majority vote and despite all the demagoguery about an unchecked Supreme Court, Congress can also pass legislation which would effectively override the Supreme Court.

But, even were a praetorian guard to be established in a majority of nations, there would still be the problem of the recalcitrant nation headed by a stubborn warmonger who would scoff at the threat of Antarctica and proceed to grab territory, expel people and otherwise make what he believes to be justifiable war. There are no justifiable wars. There surely would not be justifiable wars in an immortalist world.

Where the threat of penalty may be scorned by a military adventurer (he proceeds on his infamous way anyhow), the next peaceful threat would be total embargo and where even economic sanctions would fail, then war — war that would eliminate the war maker and not leave the miscreant in power. In an immortalist age he would be packed off, as Napoleon was packed off, first to Elba, then finally to the island of Saint Helena where he died a death deserved only by devils (stomach cancer).

Taming the American Oligarchy

America is defined by its ruling oligarchy and the definition they use is the right to earn profits wherever the opportunity for them exists. The Constitution offers no specific right to make a profit. The right to do so is implied in several articles of, and Amendments to, the Constitution. The freedoms guaranteed by the Constitution are promoted and protected by the oligarchy because, taken all together, they shelter the rights of the individual against the federal government, leaving the oligarchy free to pursue its principal interest, which is to turn a profit wherever in the world it can. Once turned, of course, there is the ever-present need to protect that profit from government "usurpation" through taxation. What is the point in earning all that money if the government can tyrannically take it away from you?

How the Oligarchy Works

The key to understanding how this works is the Republican Party's determination to control the federal courts, beginning with the Supreme Court. Indeed, the US Supreme Court is, when there's a conservative majority, the praetorian guard of the Republican Party, which is to say American capitalism. Naturally, the aim of the Republicans is to appoint as many federal judges as possible during those occasions when they are in power; and federal judges are appointed for life, not for four-, six- or even ten-year terms. It works exactly the same way for Democrats. Never was this more apparent than during Franklin Delano Roosevelt's second administration (1936–1940) when FDR, frustrated by the Supreme Court's unwillingness to give constitutional blessing to his anti-oligarchic New Deal, was tempted to "pack the Court." The attempt brought down the national roof (and earned the enmity of FDR's then Vice President, John Nance Garner). Roosevelt learned, painfully, that some things about the American political system are not subject to revolution without bloodshed and the US Supreme Court is one of them. So the US Supreme Court is not only the arbiter of what is or is not constitutional, it is a sacred cow.

Is Market Exploitation of the Chinese Aggrandizement?

China's future never looked brighter. Its policies regarding population control, critical to an immortalist society, have already placed it in the forefront of nations. China also appears to understand that its allure for the American and European oligarchy is the prospect of marketing to a billion industrious and intelligent people. Hong Kong, Macao, Taiwan — none is worth demolishing China's status as an emerging leader of the new age.

China is no fool; it is well aware that Western oligarchs are rubbing their hands over the prospect of serving the billion-plus Chinese economy and it will demand *quid pro quo* as it gradually opens its doors. The question for us is whether or not the American and European marketing plans for China properly belong under the heading of aggrandizement or under acceptable economic behavior. Aggrandizement carries the stigma of immoral behavior, of benefiting oneself at another's expense, so the judgment must not be arrived at lightly. To condemn all economic expansion as aggrandizement would bring the economies of the world to a halt. Provided that there is a clearly recognizable benefit to the consumer, there is no moral wrong in economic expansion.

The invention of toilet paper, sliced bread, frozen foods, the automobile, the airplane, the telegraph, the telephone, the steamship and the thousands of other bona fide products invented by creative individuals have benefited and still benefit all mankind, even if a yearning for personal wealth, fame and historical immortality was high on the list of motivations. It would be sinful if beneficial products were not offered to everyone in the world, so normal economic expansion, and the human ambition which drives it, should not be properly counted as greed unless we would be willing to count all profit-seeking and profit-making as greed, which would, as noted, bring the world's economies to a standstill.

Desire and Ambition versus Greed

Mortal or immortal, a distinction must be drawn between desire and ambition on the one hand and greed on the other. In economics, greed amounts to what drove the original Rockefeller to corner the supply of kerosene — for no other purpose than to increase its price. ("I believe God gave me the right to make money and more money and then to dispense as I will.") Well, God did act through the United States Government which legislated the Sherman Anti-Trust Act, effectively breaking up the Standard Oil Company (Rockefeller's trust), as it more recently broke up the American Telephone & Telegraph Company and as it threatens to break up the Microsoft Corporation.

In business, greed is also the attempt to create gigantic horizontal enterprises for no other purpose than aggrandizement. The General Electric Company is an outstanding example. Since the various enterprises gathered in a horizontal conglomerate (what do RCA and NBC have to do with light bulbs?) do not serve each other productively, they really have no other purpose than to get bigger and bigger (naked aggrandizement). The rationale for a horizontal conglomerate is that the companies thus acquired would yield a greater return (profit) than the same capital invested in stocks and bonds or left in the bank would. Indeed, a concomitant purpose of forming a horizontal conglomerate is precisely to become the sort of stock in which financiers and speculators will invest their money, thereby bidding up the pre of the stock and thereby further enriching the principal shareholders and the heads of the conglomerates, who are these days paid with options on huge amounts of the conglomerate's stock. What else can one call such behavior except greed?

David and Bathsheba

During the reign of King David, a popular couple named Uriah and Bathsheba traveled to Jerusalem to see the great city built by David. Bathsheba was an extraordinary beauty and her husband was everywhere hailed as a gifted general. One morning, while taking a breath of fresh air on her balcony, Bathsheba was spied by David, who was immediately mesmerized by her beauty. How would he ever win this splendid woman, thought David? Then a devious and evil thought occurred to him. He rushed back to his court and instructed his privy counsel to offer Uriah the post of commanding general in a war that was not going well. David, usually portrayed as a brave and decent fellow, had turned scoundrel for the love of a woman he had seen in the flash of a moment. Obviously, David hoped that Uriah would be killed in battle, and indeed Uriah fell to an enemy's sword. Uriah's death cleared the field for David. Shortly thereafter, he and Bathsheba were married. Their marriage produced a son, Solomon, who grew up to be rich beyond imagining and wise beyond his years.

Solomon married a pharaoh's daughter and from that beginning married the daughter of every king with whom he signed a treaty. At the peak of his powers he maintained a harem of 700 wives and 300 concubines.

There are many reasons why kings, sheiks and tribal chiefs maintained harems – as a diplomatic gesture, or to compensate for a shortage of men lost in battle; or simply to acquire a beautiful being, sold by her parents into sexual slavery — but the overriding motive would appear to be lust. We need only look into our evolutionary history to see the naked truth.

Diane Fosse describes the silver-back gorilla of Central Africa as a peaceful fellow who scurries back and forth in the tall grass, thumping his chest and bellowing threats at a sexual interloper, rather than fighting him. Which is not to say he refuses to fight at all; gorilla skulls bearing deep tooth marks attest to the silver-back's readiness to do battle when all the whoop and holler fails.

His cousin, the chimpanzee, follows a different strategy; he includes lieutenants in his entourage who will fight by his side and whom he rewards with an avuncular pat on the backside, sending him in to enjoy the favors of the dominant males' wives.

Money Is Rarely a Motive in Rape

But what does money have to do with rape–murder? Not very much. Where rape is incited by sexual desire, money is not the motive. Sexual desire is. But sexual desire prompted purely by sexual desire falls under the heading of "over-reaching" behavior: avidly desiring a woman who wouldn't be caught dead making love to you. Yes, rape is almost always a crime committed by men, which gives some credence to the belief that the essential motive in rape is power: the need to subjugate women and to draw a lust-filled sense of domination over them. Rape following military conquest would clearly appear to be motivated by a will for power, to which must be added the will to inflict the ultimate humiliation on men who could defend neither their country nor their women. But even where rape appears to be inspired wholly by sexual impulses, we find that the rapist is a multiple rapist and rape therefore a crime incited by aggrandizement (a non-Lothario burning to have what the true Lothario has: universal sexual appeal) and therefore greed.

Does Greed Also Motivate Murder For Revenge?

Are we to conclude that greed is also the motive compelling murder for revenge? Yes. What else is vengeance but the need to equate one killing by another or to raise the game a notch by killing someone who bilked you out of eighty or a hundred thousand or a million dollars? Or stole your patent or copyright? Or raped and killed your wife and daughter? Vengeance is the exercise of retributive power and is properly classified in the compulsion to aggrandize form.

Will long extended life spans solve the problem of rape and sexual jealousy? The exchange of the hurry-up-you-must-get-it-done-before-you-die mentality for the what's-your-rush-you've-got-a-thousand-years-to-get-it-done should certainly release the sexual pressure which explodes into rape and perhaps even prevent its accumulation. Changes in the blue-nose laws and attitudes regarding prostitution would also help.

The Compulsion to Reproduce

The compulsion to reproduce almost instantly introduces the riddle of whether we are compelled to make love in order to get us to reproduce or whether or not we are compelled to make love simply to enjoy the pleasures of love-making. Freud was fairly certain that two separate mechanisms were at work: one which draws us to sex for reproduction, the other for sexual pleasure alone. It would be fair to say that biologists do not share Freud's view, believing instead that sexual attraction, the compelling prospect of it, is what gets us to reproduce.

In biology, following Darwin not Freud, sexual selection is a corollary to the axiom of natural selection. Many biologists equate the two, which is to say that the characteristics which attract us sexually are the characteristics which will persist in future generations. *Sexual selection = natural selection.* Where women are the sexual selectors women, in effect, determine the course of natural selection. Where parents are the selectors, then they set the course of natural selection. The same where men are the choosers, as in primitive tribes and in Western culture. (Has he proposed yet? Has he given her an engagement ring?) What the equation of *sexual selection = natural selection* says is that, where we think we're choosing sexual partners for the sheer pleasures of sex, we're really making unconscious choices about which characteristics (read genes) we would like to see persist: tallness, strength, high intelligence, leadership, money-making ability, large penises in men, beauty, shapeliness, intelligence, loyalty, and small vaginas in women.

What biology suggests is that there is one fundamental drive motivating sexual behavior and that is the compulsion to reproduce, the irresistible pleasures of sex being the means by which our genes induce us to reproduce. The principle is never so obvious as in the rest of the animal world where estrus ("heat") is the signal inviting sex. Where there is no signal, there usually is no sex. Natural selection has favored human beings with a constant desire for sex, so much so that sexual pleasure has taken on a seeming dimension of its own — which is a lovely delusion.

The logical proof that sex for pleasure springs from the same mechanism as sex for reproduction lies in the most obvious set of facts. In normal behavior, sex for pleasure requires a penis and a vagina. It also requires penetration of the vagina by the penis. In healthy males, orgasm is accompanied by ejaculation of sperm which, in the absence of ovulation or the presence of some contraceptive device, is wasted. The mechanism is the same in sex for reproduction except that the ejaculate is not wasted. The result is impregnation. The only question remaining is which came first: sex for pleasure or sex for reproduction, and the answer is sex for reproduction. In evolution, the variations which produced us were mutations acting on the genes of a proto-human. Our final appearance as *Homo sapiens* was the result of sex for reproduction in which the final form of *Homo sapiens* emerged. So, when we ask which came first, the chicken or the egg, the answer is always the egg (bearing the final mutations which emerge as the chicken).

What we seem to do in life, either because we are bored or because we need to gain distinction before we die (read fame and glory on the way to historical immortality), is to make metaphors out of humdrum events. Sex for reproduction is distinguished from sex for pleasure because pleasure is more often than not the reason why we seek sex. Why? Because it is perhaps the most pleasurable of human activities and it is the most pleasurable because evolution has favored individuals with stronger and stronger sex drives.

The same might be said of the distinction between appetite and hunger. Taste buds enhance our enjoyment of food, which induces us to eat and thereby be nourished, and so we make a fetish of the most appetizing foods (gourmet cooking, exotic foods, highly imaginative chefs). We reach the point where the mere thought that we might be eating gourmet food because we're hungry is wholly and totally déclassé and may result in permanent cancellation of one's French visa. We do the same with clothing, reaching for the sometimes absurd peaks of high fashion. The modern architectural maxim that form follows function is, when it is followed, a rare example of realism in one of the arts where surpassing reality, or revealing what we never see except through the eyes of a visual genius, is usually the rule. After all, where in nature can one enjoy the titillation of a warped wrist watch or a deformed goddess descending a staircase in the nude?

Why the compulsion to reproduce? Because it is one of the things the atom does: it reproduces itself. By reproducing itself I mean that the atom gathers in aggregates which it then strives to maintain. One of the ways the atom maintains its aggregates is to gather other atoms into aggregates similar to or exactly like itself. By the statement "the atom reproduces itself" I do not mean that the atom creates more atoms. To create more atoms where they did not exist before would be tantamount to making something out of nothing, which the atom, mighty though it may be, cannot do. Nor can the atom become nothing; neither can it become something else.

One may thumb one's nose at the compulsiveness of reproduction, or think one is thumbing one's nose, but as one expires and his remains (read atoms) are tossed to the winds or fed to appreciative maggots one must claim with his dying breath that all the things he has accomplished in his puny life are worth more than the most awesome phenomenon in the universe: reproduction. It's worse than that; for as one smugly ends a childless life one must acknowledge that what one has really thumbed one's nose at is the millions of generations it took to make one human being, one giraffe, one whale, one lowly snail. For unless one's forebears were alive at the time you and they and the countless generations before them were alive, you would not be here. (See above.) The sex cells you carry in your loins have never died; they and you have been immortal until now.

But what is that compared to the need or desire to be rich and famous? Why must one reproduce oneself just because one has been immortal until now? Because everything else you have achieved is worthless compared to the unfathomable joy of reproduction. Nonsense, you might say. What is a baby compared to a million dollars? A million dollars means freedom from work, a life of leisure, a sense of accomplishment. A baby may initially bring joy but unless you're already rich, a baby is a burden shackling you to eighteen years or more of work and care. But while a million dollars may buy one freedom from work, a million dollars multiplied a thousand times over will not match the wondrous excitement of watching the miracle of a son or daughter walking, laughing, talking, and preparing to reproduce himself or herself in the endless procession of life.

It may be, as a song suggests, that this is all there is — which is to confirm the opinion of many physicists and one misguided poet, all of whom were/are certain that the universe will go out not with a bang but with a whimper. Then again, it may be, as noted, that evolution on its own may finally produce a species that does not die. It may be, too, that all living things will one day live forever or be relatively immortal. It also may be that the entire process of expansion and contraction of the universe has repeated itself an infinite number of times before and will repeat itself an infinite number of times in the future. That is the major implication of a universe that could not have popped into being once, and then exploded into the stunning spectacle which it is.

The Evidence of Evolution

All this philosophical speculation about immortality, or relative immortality, maybe yes, maybe no, does not gainsay the fact that we still grow old and we die, and sometimes we die before we grow old. Facts, hard facts. But unless one is ready to throw Darwin and all of medical science out of the window the facts are that we slowly, torturously slowly, have evolved from a single cell organism to the awesomely complex creatures we are without missing a stroke. Millions of organisms have spun off that single cell manifestation only to fail and become extinct, but not us; not any of the plant and animal forms that abound on the earth today. We have never missed a stroke. We have adapted to thousands of everyday mutations and we have evolved through fish, reptile and two-legged apes to the marvelously divine but equally diabolical creatures we are without missing a generation.

On the surface it would seem that evolution is at odds with the compulsion to maintain form, but beneath the surface there is the evidence of the old reptilian brain, of the heart, lungs, kidney, blood, stomach, digestive juices, waste disposal system, reproductive system which we have in common with apes, cows, dogs, hogs, sheep, whales and other mammals (each of which species reflects successful adaptation of mutations to environment). Rather than see the proverbial bottle half empty, we would do well to see it as half full, which is to say that beneath the seeming differences between human beings and other animals there are the similarities which remind us that we share a common evolutionary development of biological systems whose purposes are nothing more and nothing less than keeping the organism breathing, eating, fleeing, sleeping — whatever it takes to stay alive and to keep staying alive — no matter what the surface appearance may be.

In effect, we will adapt to whatever mutation permits us to maintain form, from a four-chambered heart to a six-chambered stomach. We are, as all living things are, in a process of constant evolution in which the sun acts upon our sex cells to modify the forms we find ourselves in without destroying them. What seems strange is that animals share similar systems for staying alive, while plant life shares another. Why would all animals have hearts, gills/lungs, digestive systems, unless there are patterns of biology innate to nature from time immemorial, through each of the infinite number of expansions and contractions of the universe, which is to say that the biological evolution we observe has repeated itself over and over, in each cycle of expansion and contraction *ad infinitum*. Unless we believe that evolution will suddenly come to a halt tomorrow or next year or in a thousand years, we have little choice but to believe that evolution is headed toward more of the same, that is, to increase the complexity of plants and animals as it has increased their complexity from single cells to multibillion cell organism, until the process results in animals who may be killed but who will not automatically die.

Chapter 9. Atoms Think (Contribution to a Theory of Quantum Psychology)

What better argument to induce us into believing in Plato's God, if not the God of the Bible and the Koran, than the process by which the atom reproduces itself? A mind would certainly seem to be at work in the fastidious way the double helix uncoils, makes a pattern of its uncoiled strand, and then, drawing only on the elements in the nucleus, forms a new strand called messenger RNA which then leaves the nucleus, enters the cytoplasm and there, in assembly-line fashion, selects the amino acids its instructions tell it to; presto, magic, another cell is formed. The atom has reproduced itself. The process certainly seems to demand a high intelligence, higher by far than ours. What part could we have had in the formation of the original cell? None, if what one means by "we" is a mind/self other than the mind/brain constituted by the atom. What part did we and do we play in the formation of the double helix and in the precise intelligence with which it forms new cells every day? None.

We have already observed that the co-existence of a mind/mind (a mind which is an aspect of a soul or spirit) with a mere brain requires of their interaction that something become something else, and we have seen why something cannot become something else (see "Big Bangism" above). Either we accept that what we call objects are ideas, or we accept that what we call objects exist independently of us and that the universe is what everyday common sense — and science — says it is: matter, pure and simple. If there be no mind then what is it that thinks? Our brains, yes. But ultimately it is the atom, the complex aggregates of which (cells) in the brain are called neurons. It seems preposterous that inanimate things like atoms could be capable of thinking; of logic; of high and low mathematics; of high and low art; but such is the inevitable conclusion of the principle of continuity and what we also commonsensically accept as the material world. The truth of the matter is that the thinking atom is not as lowly as we would like to think. It knows more about our bodies, for example, than our vaunted "minds" do, even if we have soul/minds. It also knows how our bodies work; what elements are required to make them work; where those elements are to be found (in sunshine, plants, and other animals).

Little By Little, We Learn What the Brain Has Known for Millennia

For example, not until the 17th century were "we" aware of the fact that blood circulates in our bodies or that the heart is the pump which controls circulation; nor were we aware that red blood cells carry oxygen to our organs, without which we would quickly die. Neither were we aware, until the 19th century, of the existence of bacteria, the airborne strains of which served to spoil wine and beer while the myriad other strains caused everything from diphtheria to tuberculosis to pneumonia. While we were thus being laid low, the lowly atom was developing an immune system with a complex and ingenious system for fighting off infection while with equal wonder it formed antibodies to protect us against future infection by the same bacterium. We were still using leeches (to draw out diseases perceived as bad blood) as recently as the early 19th century. Penicillin and antibiotics were discoveries of the mid-20th century; aspirin was developed in the late 1890s. Secular medicine could not really be said to have begun until the time of the ancient Greeks (Hippocrates 460–372 BC) and Galen

(AD 129–216), a Greek-speaking Turk, though the ancient Egyptians did leave a body of medical lore.

Truth be told, "we" were totally ignorant about what really goes on in our bodies until the early 1600s (dissection was forbidden in the ancient world) and we had been around with mind/minds since the fabled Garden of Eden — or around without "minds" but with developing brains for nearly 3 million years. In spite of our ignorance we survived, evolved, and developed, and we did so because the lowly atom is infinitely more intelligent than "we" are and was on guard to maintain the form of the aggregate it found itself in; constantly attentive to the growing complexity of our bodies; vigilantly adapting to the bombardments of mutations, favorable and unfavorable. When a mutation is not favorable, in effect when a healthy gene is damaged or when a gene is ill-formed, we are burdened as we all are with a recessive gene, recessive because a healthier version of itself rides up front while the recessive gene lurks in the attic like a demented relative hidden away from the neighbors and suitors. Until. Until, as bad luck would have it, we witlessly mate with a carrier of a similar recessive gene and the embryo or fetus self-aborts. Or the fetus develops normally and grows into a seemingly healthy individual who dies a sudden death at forty from diabetic coma or ventricular fibrillation. Our brain atoms don't look so smart then, but the hard truth is that they did all that the life process obliges them to do — they hid the recessive gene until it found its mate; then they could do no more. Or it may be that they also had the additional task of getting rid of a pair of recessive genes by throwing them back into the pot, scrambling them and dispersing them. It is as though the atom forges blindly ahead creating an aggregate, then relentlessly maintaining and aggrandizing that aggregate (the form it created and finds itself in) until it is overcome by a more powerful aggregate (a large man kills a smaller one; a big fish eats a little fish; the sun mutates a gene); all the while knitting and compiling beneficial mutations until the damn thing is perfect; perfect in the sense that it does not self-destruct (die). It takes a while.

Why Is Nature So Slow?

If our brain atoms (neurons) are so bright, why are we dying in droves? Wrong question. The proper question: Why is nature so slow? If nature had a viewpoint, it might be that it is moving relatively quickly. It takes billions of years for a planet like ours to form and another billion or so for atoms to enjoy the relative peace and quiet required to form simple biological aggregates, another billion or so for biological aggregates to form extremely complex aggregates, and another billion to reveal what has been formed and, perhaps, what is being formed if we would only look hard at the evidence. Anyway, why the rush? Well, we don't want to miss the parade. We want to be among the marchers when the first human being celebrates her two hundredth birthday; then again when she celebrates her thousandth year of life. Or, better still, we want to be among the lucky ones drinking to the replacement of our worn out telomeres. We don't want to die, not yet. Or, if we must die of a heart attack or cancer, let it be with the knowledge that we collapsed on the way to our six hundredth birthday party. If we were as smart as we think we are, we would know why we die of cancer and heart disease and we would know what it takes to forestall such untimely deaths. If we were as smart as we think we are, we'd stop killing people in war not only because it is the most callous of crimes but because, among the uncounted millions of war dead, there might have been a thousand or more potential centenarians blessed with the mutations favorable to a protracted life span, men and women who might have met their genetic matches and moved us that much closer to what appears to be the end point of evolution: biological immortality.

Bad Luck Separates the Losers from the Winners

Truth be told, we continue to die of cancers, heart disease and innumerable other ailments without really knowing why except to blame diets, smoking, breathing coal dust or toxic chemicals. All we can be sure of is that natural selection is doing its snail-like work winnowing out the losers and revealing the winners in the process of beneficial and not so beneficial hits. One person dies of a diabetic coma at forty, another lives to one hundred and twenty-two reporting that she was rarely ever ill. Even now that we believe longevity is a matter of shrinking telomeres, we don't know what relationship, if any, exists between our telomeres and freedom from illness. The heroic cell biologists who solved the deadly problem of disappearing telomeres cannot be reproached for solving the problem of so-called natural death before they solved the problems of heart disease, cancer, Alzheimer's, Parkinson's, cystic fibrosis, malaria, paraplegia, blindness, deafness, and muteness. It was always a possibility. Our "old souls" might have warned us; might have told us how to achieve the proper sequence. But they didn't know. They still don't.

Bittersweet? Perhaps. But look at the afflicted and you will probably notice them smiling and guzzling champagne with the rest of us. By then, we surely will have learned to trust our instincts before we trust our conscious minds, seeing through our myopic eyes only the preponderant evidence of our mortality. After all, how smart can our conscious minds really be when they can see no better than elephants?

Elephants Bellow for Justice. Why Punish Them for Eve?

Elephants mourn the death of a fellow herd mate, standing around the fallen behemoth bellowing and wailing his or her inability to rise, run, and join in the bellowing. Elephants mourn their dead in funerals in which they move the corpse to an elephant graveyard. They cry out to the heavens for explanation, for correction of the injustice, for some sign that death is an error, that it won't happen to them, because every silent electro encephalic wave assures them that death is wrong; that it shouldn't happen, and certainly shouldn't happen to him or her. Like us, like every living thing, the elephant's gametes have never experienced death; they have evolved over millions of years without missing a stroke, each generation of its parents alive and happy when it reproduced. The elephant who falls for the first and last time never believed he or she would be anything other than what it is in its prime; unless it were unlucky to have been present at a funeral of a mother, father, grandparent or an unfortunate buddy brought down by a poisoned dart for food or, more likely, a treasure trove of ivory. After all, can a piano reproduce the sound of a Beethoven sonata with a plastic keyboard? Heavens.

Why Teenagers Think They're Immortal

We sneer when human teenagers speed along dark, curving roads, under (or not under) the influence of alcohol or drugs, challenging something, dare deviling death, hoping to impress comrades by their courage, the more foolhardy the better. We say, barely hiding our contempt, that they think they're immortal. But they have been immortal until now that their derring-do has done them in. Nothing in the promise of relative immortality can be counted on to prevent accidental death even as the sense that we are immortal now cannot forestall a fatal automobile crash. Indeed, the innate knowledge that we are immortal may lead us to bravura performances that end in death. So we hear mothers and fathers shouting warnings about the dangers of being maimed or, worse, being killed. Children sneer. What do chicken-hearted, bird-brained parents know about it? Children aren't stupid. They are not suicidal. They are simply responding to the certainty of our neurons that they are immortal: as individual atoms and as the complex aggregates of atoms of their brains. Our brains have never experienced death. What we experience is someone else's death, far, far away from our own, thank God, and then only as a witness to what everyone assures us is death: a cold, immobile, silent corpse whose soul has flitted to another place. Or will be trapped in the grave with our bodies, maybe yes, maybe no, for eternity. Who knows? Who knows, indeed. No one tells us where our souls are parked while they await the burial of our bodies. A small detail some theologian will have to cover. For the most part, a billion Roman Catholics, another billion and a half Mohammedans, nearly a billion Orthodox Christians, a billion Hindus and Lord knows how many protestant Buddhists believe the soul migrates immediately upon death of the body either to heaven, purgatory, or hell or to some holding pen awaiting reincarnation on the Day of Judgment. Hindus await a second birth, Buddhists await the long term re-aggregation of our original atoms, Christians anticipate the re-union of their souls with a body, perhaps the same one, perhaps not. Why does it matter? Plato believed that the body, like all material things, is a phenomenon; basically unreal. Our neurons pretend to accept these various strategies because they seem to affirm what our neurons really believe which is that they are immortal.

The Making of a Daredevil

What is a young child to believe? Commanded to believe in life after death, he accepts the thought reluctantly, all the while cringing as he kneels to pray at the bier of a classmate who drowned. So it's true, he or she thinks. You can die. But what the children mean is that other people can die. They cry for the deceased because they, the dead, have been deprived of the joy of being alive or because they pretend to accept that a fallen comrade will not show up at school tomorrow, nor appear to play ever again — forever. But they don't believe that it will happen to them. So a boy grows up to become a daredevil, risking his life at high diving, sky diving, parachute jumping, bungee jumping, mountain climbing, hang gliding, racing in competition, motor bike racing, drag racing. He does it to impress the girls who, responding to primitive urges, allow him to believe that absolute fearlessness, not money, power, intelligence or longevity, is the real measure of a man. The girls copycat by allowing themselves to be persuaded that they are equal to the challenges which men undertake and they are so persuaded because they believe they must prove their equality, for only by such proof will they merit the privileges that men enjoy: fame, riches and power. Poor dears. They don't understand that men do it to win women, fame and riches.

By the time we reach adulthood, we are at least playing the odds: we don't believe we're going to die but why take a chance? The doomsayers keep pointing to the mounds of corpses accumulated over the ages to remind us that death is inevitable; only a fool would believe otherwise. Gradually we come to the churlish acceptance that we had better buy plenty of life insurance, make a will, provide for our children, take out a policy on the wife in case she pre-deceases us and we have to hire a housekeeper to replace her. Closer to the heart, we also come to accept that we are doomed to make "the supreme sacrifice" of being killed in battle. We coin fatuous, fatalistic shibboleths like believing that there is a bullet with our name on it, and if there is not then we will emerge from battle unscathed. Corpse by corpse, the doomsayers and the deranged build one death culture after another — except in sane, civilized ancient Egypt, whose slogan might have been "never to die."

Making Up for Lost Time

All right, we say, so it has taken us thousands of years to begin to understand the process of what truly goes on under our skins, but clearly we are making up for lost time. We are indeed. But we have merely scratched the surface. The discovery of the double helix and the establishment of molecular biology have already begun to reveal how organs diseased by defective genes may benefit by gene replacement therapy; the discovery of stem cells which may serve to regenerate organs; and the discovery of telomeres and telomerase which may be the long-hunted secret of death — these constitute a revolution in biology rivaling the discovery of the microbial world. What we need to do now is to persuade the Trekkies that the exploration of space is premature and that those resources would be best applied to the total revelation of the "small world," both inside and outside of us.

Now that cell biologists have solved the problem of "natural death" and with it, we trust, the problem of cancer, we may find that the need to protect ourselves against the missile-might of rogue nations is more an exercise in pork-barreling for the military-industrial complex than it is a scientific exercise in military preparedness. It is an unconscionable waste of money and talent to build a sky-crossed missile defense against the likes of Iran, Iraq, and North Korea. They must be made to understand the horrors of missile retaliation. It is stupid to believe that any alliance of small nations would dare challenge the superpower alliance of Europe, America and Asia.

Look At All We've Created

And so we are left with the embarrassing evidence that we are not as intelligent as the unconscious; we are incapable of devising systems as brilliantly conceived and engineered as life. Well, we say, we have discovered a long list of complex systems: telephone, computers, radio, television, airplanes, rockets, the A bomb, the H bomb, the thousands of medicines that bring relief, if not cures, from both the everyday and the extraordinary maladies that afflict us. But we have not as yet created life. We have not created what dumb atoms were able to create in reproducing the cell, nor built on that seemingly miraculous achievement a system like the double helix to collect and preserve on its twisting ladder the elements (A, C, T, G) that distinguish the aggregation as a distinctive reproducing organism, then preserve the mutations that enabled it to maintain form — even when the mutation deformed the organism.

It is a spectacular creation, worthy either of a God or of a universe consisting of matter which thinks which is why we think; a creation evolving in the way it has evolved an infinite number of times in an infinite number of expansions and contractions on an infinite number of planets: the same mountains, the same oceans, the same trees and plants, the same animals, the same people all headed for a multibillion-year journey, not into the unknown but to the known and the predictable, to the maintenance of the forms we find ourselves in (forever, if possible, but if not forever then for as close to forever as our competitors or the dinosaur's fate will allow).

We are, or are about to become, what we have projected into our gods: immortal, unconquerable, kind, just, wise and loving. Of course, we think we're gods already, fantasizing about Star Treks and Battleships Galactica while we sneeze our ways up the gangplanks, stopping to make certain we've packed enough Nyquil and tissues. We may one day, billions of years from now, if our sun ever burns out, have need to migrate, to find new planets to colonize, but the notion that the incomprehensible vastness of the universe may be conquered ("the conquest of space") is indeed pure fantasy — infinitely more fantasy than the possibility that we might live for ten thousand years. There was a time, not too far back, when we seemed to be all thumbs, back before Watson and Crick, when there seemed to be purpose in traveling at least through our own galaxy hoping to discover intelligent beings like ourselves, who might be a hundred thousand or a million years ahead of us, who might share their inventions with us. It's still a worthy goal but not worth the priority we give it. Time enough for space travel when we're living a thousand years or more, when we might be interested in whether or not there are problems to be faced at various levels of longevity: how big can we expect to be on our 500th birthdays (there is a law of increasing size in biology); will we have lost the ability to reproduce ourselves (should we or should we not clone)?

Trusting the Unconscious

Until then, we need to regain the trust we once placed in the unconscious which, truth be told, did not always guide us wisely in the past. We listened too closely to the neurons' call for immortality and, faced with the overwhelming evidence that we all must die, we invented gods, heavens, and immortal souls to resolve the conflict (again, except for Buddha and Judaism). Was it absolutely wrong? No, nor was there a choice. Better to believe that life moves from one plane to another than deny the yearning for immortality which, at the unconscious level, amounts to certainty. It was also a mistake to trust the unconscious to reveal natural law for as long as we did. But Christians had their own guides, namely Plato and Aristotle, and it took half a millennium to accept that scientific method, not Platonic ideals, was the way to discover and understand the world in and around us. However, unless we are willing to credit a bolt out of the blue or divine intervention the decision to rely on scientific method rather than pure hunch has to have been inspired by the unconscious. Not that scientific method is free of hunches. To the contrary, scientific method always begins with a hunch, an Archimedean *Eureka!* which is then put to the tests of trial and error. Everything we will ever need to know is now known to the unconscious. It knows when something awful goes awry with our bodies. It knows when a cancerous tumor has formed. It tries to undo the damage by various deployments of our immune system. It desperately cries for help when the tumor assaults our nervous systems, then meekly or angrily, but always reluctantly, it admits defeat and dies. It has failed the compulsion to maintain the form of its organism. It has also failed to communicate to our consciousness what was going awry while it was going awry.

We need to stop treating our atoms as though they were just tiny aggregates of inanimate energy and start treating them as what they are: thinking units of being, who know exactly what they are, where they should be. (A free electron flitting about searching for a home in another atom's orbit and disturbing the peace of the universe was once microbiologists' favorite cause of death; they need to hold on to that thought; we may save our telomeres and find that free electrons are still the bad guys.) As a thinking unit of being, we need to treat the atom as a reasoning creature, asking it every question we can think of, then observing its answer or listening carefully to how it responds. After all, since the atom has only three compulsions to motivate it, how complicated can the answer be? For example, one aspect of cancerous tumors that oncologists seem not to have given much attention to is the reason for the immortality of cancerous cells (see above). We know the philosophical answer, but philosophy will not discover a cure for cancer.

The Atom Is an Extraordinary Thinking Machine

Well, we think it is outrageous to suggest that atoms are thinking entities. The very least we should expect is that atoms cannot think until they have formed the complex aggregate of the brain. After all, a bone may be a complex aggregate of atoms but a bone cannot think. My, my, what heresy. A bone is a brilliant thinking machine, every cell of which is chock full of DNA constantly producing red blood cells to carry oxygen to every organ of the body. A bone's DNA also knows when and how to knit itself together again, albeit sometimes crookedly so it needed the astuteness of the conscious "we" to observe that a plaster cast might help the bone to knit correctly. So the atom thinks and what it thinks about is continuity; because it is such a remarkable thinking machine it also adapts to an organism fulfilling its role as a red blood cell or an optic nerve. It also never forgets that it is infinitely continuous, a condition it imparts to the organ of which it is a constituent. What? we think, as we go blind. I have lost my sight? Impossible. I cannot be blind. Isn't there something that can be done? The unconscious, what consciousness thinks of as the unthinking part of us, has now been revealed to be the awesomely intelligent world of cell biology where feats of creativity take place which are still beyond us.

Hunches Are the Mothers of Ideas

Where we are reluctant to regain the trust of our atoms for fear that we will abandon observation, we need to be reminded of Archimedes and his shout of *Eureka*! We need also to be reminded that some of our weightiest decisions are made on hunch; falling in love being the most notable. The calculating lover may marry for money, but the man or woman who binds himself to another for money will soon find that he or she has purchased a fountain of bile. Love is the way the atom draws us to one another for the purpose of reproduction. It is an attribute of the compulsion to reproduce. It is foolhardy to attempt to analyze it. It operates purely on the level of the unconscious. We may dissect the object of our affection, putting him or her through a series of trials testing for suitability as a mate and a parent. And, our hearts heavy, we may reject the other as unsuitable for any of a dozen reasons; but our decision will not gainsay our love.

Until the Great Day Arrives, Reproduce Yourself

Unfortunately, the unconscious is relentless about its compulsion to reproduce and will not stop nagging until the job is done. We may rebuff it, choosing to remain childless, but we do so at great peril to our peace of mind and with the sense that we are somehow unfulfilled no matter how richly the world has rewarded us for talent and hard work. If you believe that historical immortality (our names on books, paintings, sheets of music; our images on film and in statues) is more than fair compensation for having lived without having reproduced, then hurrah! for you. Neither of us will ever know what remarkable boys or girls you might have given to the world; if nothing else, you might have produced a clone-like copy of yourself, allowing us to enjoy your great beauty and talent for another century. And with luck you might have produced the next Darwin, Watson, Crick, Edison, Einstein, Alexander Graham Bell or Guglielmo Marconi. We will never know. Better to remember you holed up in one of your hotels, emaciated, cadaverous, with a vagrant's long, matted hair and dirty, untrimmed fingernails.

The Unconscious Gives Us Love and Creativity

Love is only the greatest gift of the unconscious. The unconscious also gives us creativity. All of the arts — painting, music, sculpture, literature, dance, and drama begin with inspiration, though they may be modified by the skills one learns in the gardens of the Medici, at Julliard or at the Royal Ballet. Indeed, it would be fair to say that much of what is most noble in life is in and by the unconscious. Well, let's not get carried away. There are many who would argue that the practice of medicine (including dentistry) is the noblest activity of all — though diagnosis leans as heavily on hunch as it does on comparative analysis. (I once toyed with the idea of ranking the noblest activities of mankind and ended by attempting to prioritize good and evil instead. I would now rank the practice of medicine as the second noblest behind microbiological research, including the discovery of medicines.)

Death No Longer the Only Winner

Still, we hesitate to put our trust in the unconscious, for the old reptilian brain continues to foment hatred and anger, the seed pods of crime and war. (Not forgetting envy in general, sexual jealousy in particular, and greed.) Conscience, which checks anger and hatred and their compulsion to rob, maim and kill, would seem to be an attribute of consciousness, something drummed into our brains as part of the socializing process. What belies that possibility is the failure of good parenting and thousands of years of religion and ethics to prevent murder and war. The nagging conundrum is why a conscience was instilled in Abel and not Cain, why in Esau and not Jacob? Do we dare renew our trust in an unconscious which continues to urge us to kill? Yes, because we have no choice.

If we are cursed with an irrepressible killer-ape gene, then we need to pray that the prospect of trading a five, ten or twenty thousand year lifetime for ninety years on Antarctica will provide sufficient deterrence against murder. In the end, we may discover that it has not been the Freudian family triangle or a killer-ape gene which has made us murderers and warriors but rather the news, then the evidence, that we must die; that we may conquer everything in sight except death, so that death is the only winner. Until this generation.